Jose Aquimix Rodriguez

Six Sigma methodology in the primary sawmill process

Jose Aquimix Rodriguez

Six Sigma methodology in the primary sawmill process

El Palisal Agroforestry Cooperative, Yamaranguila, Intibuca, Honduras

ScienciaScripts

Imprint

Any brand names and product names mentioned in this book are subject to trademark, brand or patent protection and are trademarks or registered trademarks of their respective holders. The use of brand names, product names, common names, trade names, product descriptions etc. even without a particular marking in this work is in no way to be construed to mean that such names may be regarded as unrestricted in respect of trademark and brand protection legislation and could thus be used by anyone.

Cover image: www.ingimage.com

This book is a translation from the original published under ISBN 978-620-2-16922-6.

Publisher:
Sciencia Scripts
is a trademark of
Dodo Books Indian Ocean Ltd. and OmniScriptum S.R.L publishing group

120 High Road, East Finchley, London, N2 9ED, United Kingdom
Str. Armeneasca 28/1, office 1, Chisinau MD-2012, Republic of Moldova, Europe
Printed at: see last page
ISBN: 978-620-6-45992-7

Table of Contents :

SUMMARY

The main objective of this research was: "To evaluate yields and lumber quality before and after the application of continuous improvements in the primary sawmilling process at Cooperativa Agroforestal El Palisal, through the application of the Six Sigma methodology"; for which a detailed analysis of the lumber production process was carried out, including the production from the forest to the storage of the lumber. With the help of descriptive statistics, the causes that generated the problem of low yields and poor wood quality, performance parameters and process capacity were identified. For this purpose, tools such as: baseline, cause-effect analysis, Pareto diagram, time and motion study, quality and yield study and Six Sigma analysis were used. It was determined that the production process was defined by the following parameters: the standard production time was 783.5 minutes per production cycle, 34% of the sawn timber affected by various defects, yields of 229.7 pt/m^3 and a sigma level of 2.4.

Based on the causes of the problem and analysis of the results, a combined improvement was implemented in the primary sawmilling process and storage area. The evaluation of the implemented improvement reflected a change in the aforementioned parameters, with a standard production time of 575.6 minutes per cycle, 23.3% of the sawn lumber affected by defects, an average yield of 244.2 pt/m^3 and a sigma level of 2.7.

ABSTRACT

This research had as its main objective: "To assess yields and quality of lumber before and after the application of continuous improvements in the primary sawmill process in The Palisal Agroforestry Cooperative, through the application of the Six Sigma method". A detailed analysis of the production process of lumber was performed, from the forest to the timber storage. Using descriptive statistics, the causes that generated low yields and poor quality of wood, the performance and the capacity parameters of the process were identified. The following tools were used: Baseline, cause-effect analysis, Pareto diagram, time and motion study, study quality and yields and Six Sigma analysis. It was determined that the production process was defined by the following parameters: the standard production time was 783.5 minutes per production cycle, 34% of wood affected by several defects, yields of 229.7 pt / m3 and a sigma level of 2.4.

Based on the causes of the problem and analysis of the results a combined improvement was implemented in the process of primary wood sawing and storage area. The evaluation of the implemented improvement reflected some changes standard production cycle time of 575.6 minutes, 23.3% of wood affected by defects, an average yield of 244.2 pt / m3 and 2.7 sigma level.

DEDICATION

To our heavenly father and creator of the universe for giving me health, wisdom and filling me with his many blessings in every moment of my life.

To my mother Victoria Rodriguez, Marlen Yesenia Nolasco and my daughter Odry Marina Perez Nolasco for their dedication, love and unconditional support, being them the fundamental pillar in everything I undertake.

To José María Nolasco, Blanca Rosa Romero and Walter Enoy Romero for their unconditional support to my family and myself during my engineering studies.

To my siblings, Kevin, Emilio, Mauricio, Veronica and Nury for always being by my side, giving me their understanding and support at every moment.

To the National School of Forestry Sciences for providing me with professional training.

To my advisors, Joaquín Sánchez and Luis Bejarano, for their time and dedication and for allowing this project to be successfully completed with their knowledge.

Alejandro Castillo, M.Sc. Marco Vinicio, M.Sc. Efraín Leguía, M.Sc. Andrea Johnson, Dr. Edgar Maravi, Ing. Herminia Palacios and Ing. Wilson Guerra, members of the Catie-Finnfor project in Honduras, for their unconditional support, moral and financial support for the engineering studies.

To the members of the cooperative El Palisal de Yamaranguila, Intibucá for allowing me to develop my thesis in their community enterprise, logistical support and understanding during the research process.

To my friends and colleagues Elder Medina, Juan Luis Hernández, Cintia Yosely Banegas, Hugo Salgado, Mario Guillen, for always encouraging me and motivating me to continue with this project.

Heidi Vides, Johnny Pérez and Dr. José Alexander Elvir for their support, knowledge, patience and technical contribution in the realization of this study and to each of the people who always shared their advice and gave me encouragement to continue and complete my thesis project.

CHAPTER 1

1. INTRODUCTION

The coverage of coniferous forests in Honduras is 36.3% (1,960,511.08 ha) (ICF, 2013). Part of these forested areas are currently managed by small and medium-sized community agroforestry enterprises.

Small and medium-sized enterprises (SMEs), in general, play an important role in the economy of developing countries, through the dynamism they bring to the economic system, their contribution to employment, GDP and the increase in the innovative process (Márquez & Pérez, 2007). Due to their size, these companies have greater flexibility to adapt to changing markets and to promote innovative projects (Martínez, 2013).

We are currently living in a new economic era in which continuous improvement has become a way of life for companies. This way of managing companies allows them to remain at an adequate competitive level in terms of quality and productivity (Fraile, *et al.* 2003). The purpose of improving productivity and quality in the processes must be continuous, leading to surpluses that are reflected in the company's profits (García, 2010).

In order to evaluate processes and maintain continuous improvement as a work philosophy, forestry companies need to use tools such as Six Sigma to continuously analyze their performance under the philosophy that the customer comes first (Harry & Stewart, 1988).

2. JUSTIFICATION

Since their inception, community agroforestry projects have received technical and financial support from government institutions and international organizations. Even so, these enterprises have not been able to achieve financial sustainability because they face various problems of productivity, yields and competitiveness with other private enterprises, to such an extent that some community initiatives have been disappearing.

The agroforestry cooperative El Palisal is one of the companies that administers the site under forest management called: EJIDOS DE YAMARANGUILA of ejido tenure, with an annual allowable cut of 5,152.104 m^3 /year of pine wood (Management Plan, 2008). The company's objective is to conserve the forest and generate employment and profits for each of its members. This last objective has not been fully achieved because financial sustainability has not been achieved despite the intervention of various institutions.

According to production up to June 2015, the agroforestry cooperative El Palisal achieved yields of 210 pt/m^3 . According to Vasquez (2012), when using a Wood-Mizer sawmill, yields of 242 to 252 pt/m^3 can be achieved according to the diameter class. This indicates that the El Palisal cooperative is obtaining low yields. In addition to this problem, the company has had problems with a decrease in the value of the product from clients due to wood defects. Therefore, it is necessary to identify the causes of the problem and implement solutions.

3. OBJECTIVES

3.1 General Objective

To evaluate the yield and quality of sawn timber before and after the implementation of continuous improvements in the primary sawmilling process at the El Palisal Agroforestry Cooperative in the Municipality of Yamaranguila, Intibucá.

3.2 Specific objectives

3.2.1 Evaluate the current lumber processing process in order to define the problem in its entirety.

3.2.2   Identify the causes of the problem affecting the transformation process, in order to evaluate and implement recommendations to improve primary wood processing, following the Six Sigma methodology.

3.2.3   Compare yields and production qualities before and after the implementation of continuous improvements in primary wood processing.

3.2.4   Design a methodological scheme of continuous improvement for the primary processing of pine wood, taking into account the mechanisms for the improvement and control of the main defects identified.

4. LITERATURE REVIEW

4.1   Social forestry system

The social forestry system in Honduras is the linkage of policies, norms, criteria, strategies and procedures for the socioeconomic development of communities and groups living in or around the forests, incorporating them into forest management, integrated harvesting, industrialization, marketing and participation in the benefits derived (Forestry Law, 2007).

4.2   Sustainable forest management

Process of permanent administration of the forest resource for one or more management objectives, related to the continuous production of products and services, without deteriorating its main values, future productivity, the environment and society (ITTO, 2000). It is also known as the management of products in forest areas of private public production of relevant economic interest suitable for the cultivation and harvesting of timber, goods and environmental services, guaranteeing their quantity (Forestry Law, 2007).

4.3 Community forestry

According to Flores (*n.d.*), community forestry focuses on sustainable forest management and the multiple use of the forest, promoting the participation of the communities that live in harmony with it, aimed at improving the quality of life of human beings in an integral manner, through three strategic components: Integral human development, integral forest management and integral farm management.

4.4 Community forestry or agroforestry enterprise

It is any private productive organization, duly recognized by the State, constituted by members of a rural community, by ethnic groups, with the purpose of managing forests, forest vocation lands and other agroforestry resources located in the area of residence and direct influence of these communities, with the main purpose of satisfying their needs in a way that allows them to obtain economic income that contributes to improve the quality of life of the beneficiary population (Forestry Law, 2007).

4.5 Community management of natural resources

An alternative to this fluctuating development model revolves around the notion of community-based natural resource management, in which the natural resources present in an area are managed sustainably and productively as a community business enterprise. Although this provides obvious ecological benefits, it makes economic sense by maximizing available resources and reduces dependence on external inputs (Grundy & Breton, 1998).

4.6   Economic impacts

According to Reiche (1992) quoted by Dubón (1996), the main task of forest economics is to contribute to the adequate and efficient administration of available and scarce resources, among different ways to produce goods and services. These must satisfy the needs of society, such as products, services, environment and generate profits for producers and companies dedicated to the rational use of natural resources.

4.7   Primary wood sawmilling process

The primary wood transformation process begins with the tree in logs or roundwood, which is the raw material for industrial wood processing (Parameda, Brenes, & Figueroa, 1998). Dimensioned wood is defined as the product processed in a sawmill or industry, with no further processing than sawing, re-sawing and planing longitudinally by a standard machine, as well as cross-cutting to give the size and the planing according to the client's requirements. Primary sawmilling companies are constantly looking to improve production systems to improve productivity and lower operating costs (Brown & Bethel, 1990).

4.8   Lumber grading

According to COHDEFOR[1] (1991), all grades of wood limit the quality of each piece as a whole, but the method varies according to the class of wood, as an example:

o Grading of common dimension lumber, common timber and structural lumber should be done on all four sides to define the worst combination of defects that affect their use, whether rustic, planed or processed.

---

[1] COHDEFOR Corporación Hondureña de Desarrollo Forestal, now known as Instituto de Conservación Forestal ICF

o The requirements for the two faces of the pieces will apply to half of the continuous narrow face.

o Unplaned areas, and consequently, lack of size, are allowed in some grades.

For the same reason, wood grading is not considered an exact science, since it depends on the visual observation of each piece and the criteria of the grader. Main defects according to COHDEFOR; knots, over measurements, humidity greater than 19%, twists, gems, cracks, fungal stains, fiber deviation, burns, resin pockets, warping, rotting, among others.

Small sawmills commonly market timber without a proper grading process, which leads to low profits and the sale of timber at a standard price (Brown & Bethel, 1990).

4.9   Quality management

Quality management nowadays consists of designing, implementing, producing and assisting in service; a quality, economic, useful and possible quality product that satisfies the demand and interests of the final consumer (García, 2010).

We can say that quality management is a process of continuous improvement (Figure 1), which involves various factors that are part of the production process, based on customer requirements.

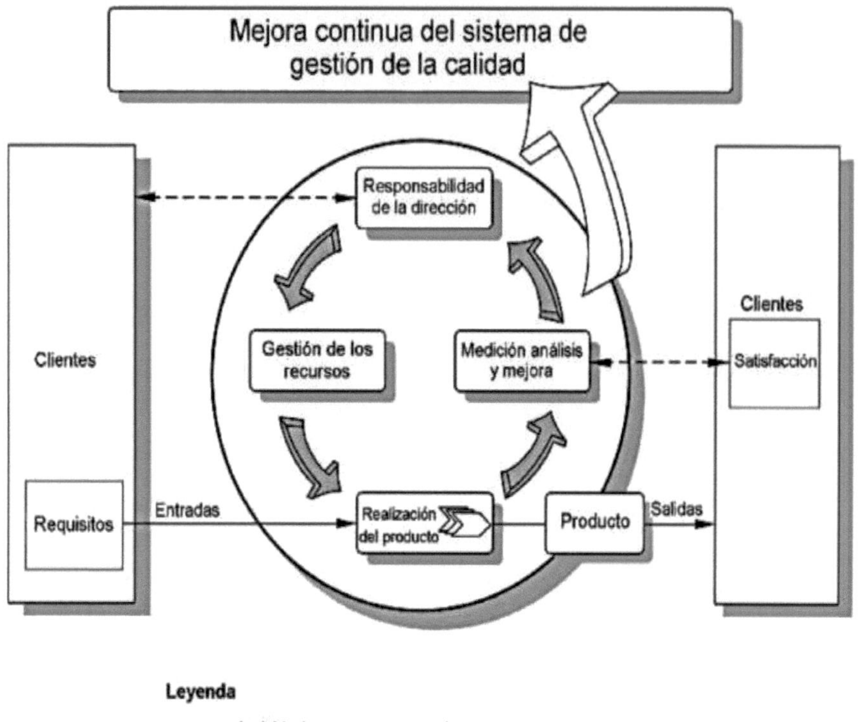

Figure 1Model of a process-based quality management system

Source: ISO 9001, 2008.

1.10 Total quality management

It is basically a business "philosophy" based on the pursuit of customer satisfaction, where

"the business process starts with the customer and ends with the customer" (Alor, *et al.* 2014).

1.11 Productivity

Productivity is the ability to achieve objectives and generate top quality responses with the

least human, physical and financial effort. It should not be confused with efficiency, which

means producing high quality goods in less time (Garcia, 2010).

1.12 Quality

It means the characteristic, group or combination of attributes that distinguish a product from the competition (Brown & Bethel, 1990). The current concept of Quality has evolved to become a form of management that introduces the concept of continuous improvement in any organization, at all levels and involves people and all processes of the organization (Alor, *et al*. 2014).

1.12.1   Total quality

According to ISO 9000 definition, "*is to comply with the customer's requirements and it is on the customer that the processes and products have an impact. The customer perceives quality in his own terms, not in those of the industry, these requirements are the ones that must be met within the production process. Meeting the requirements means; identifying the customer, taking the initiative to communicate with the customer, identifying their requirements and expectations, agreeing the requirements, writing them down and notifying the appropriate party of the agreement*".

1.12.2   Principles of total quality

Principles defined by ISO 9000:

J Quality is the key to competitiveness.

J Quality is determined by the customer.

J The production process is throughout the organization.

J The quality of products and services is the result of the quality of processes.

J The supplier is part of the process.

J Supplier-customer chains are indispensable.

J Quality is achieved by people for people.

J Establish a zero-defect mentality.

J The competitive advantage lies in error reduction and continuous improvement.

J Everyone's participation is essential (collective conscience).

J Quality is first and foremost a management responsibility.

Quality improvement is related to productivity, efficiency and the customer (Figure 2). When this synergy exists within the production process, a better competitiveness is achieved and, consequently, the company remains in the market (Deming, 1986).

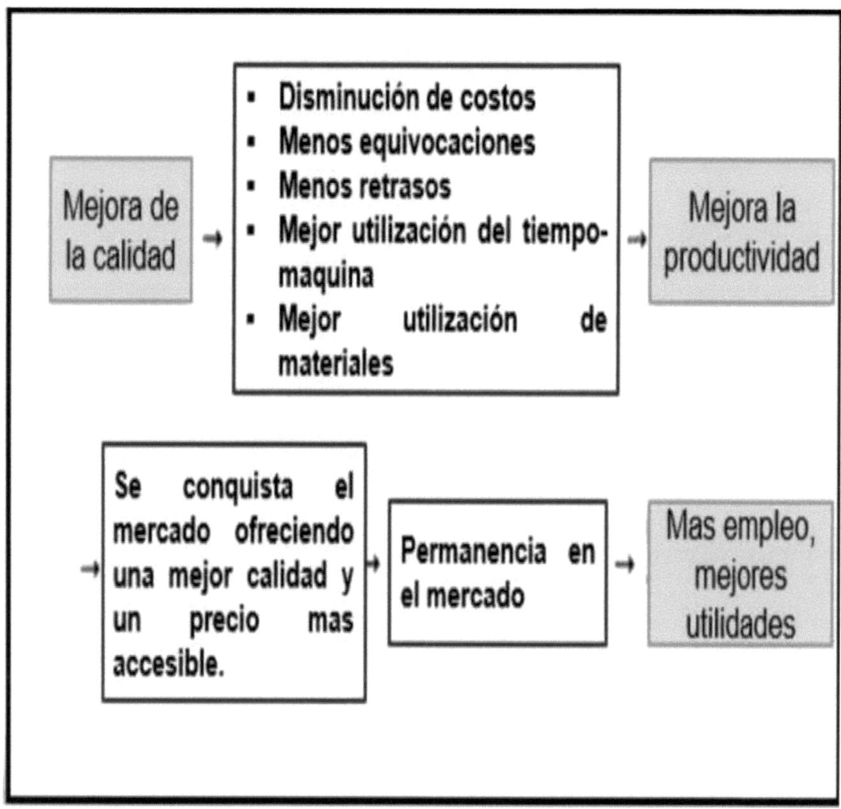

Figure 2 Quality, productivity and competitiveness
Source: Deming, 1986.

2.13 Continuous improvement

It is a productivity enhancement tool that leads to constant and permanent growth in all phases of a process. Continuous improvement ensures the permanence of the process and the probability of improvement. Some of the tools used include corrective actions, preventive actions and customer satisfaction analysis. It is the surest way to improve quality and efficiency in organizations (Garcia, 2010).

2.14 Six Sigma" Methodology
2.14.1 Background

In the 1980s and early 1990s, Motorola first introduced the Six Sigma initiative, which led to a reduction in the number of defects in its products from 4 to 5.5 sigmas, resulting in $2.2 billion in savings. The second in applying the Six Sigma methodology was the General Electric Company, which indicated that it could increase savings by 50% more in five years, if used successfully (Fraile, *et al.* 2003).

2.14.2 What is the Six Sigma methodology?

"It is a quality philosophy based on the assignment of achievable short-term goals focused on long-term objectives". This methodology seeks to reduce process variation, defects and errors in all processes of a company in order to increase the percentage of market demand, reduce costs and increase profits (Fraile, *et al.* 2003).

The use of Six Sigma starts from knowing the background to identify the losses proportional to the variability of the quality characteristics of the company's product. The fundamental strategy for quality improvement is to identify the causes or factors that produce the problem and then correct it, so that there is a minimum variability (Barbosa & Chancusig, 2012).

2.14.3 Phases of the Six Sigma methodology

The DMAIC procedure[2] (Figure 3), is one of the procedures of the Six Sigma methodology and due to its nature it is applied to companies in operation, with the objective of improving already existing processes (Pries, 2006).

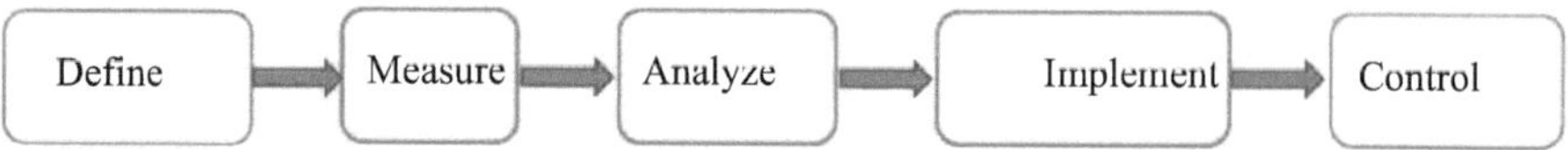

Figure 3 DMAIC diagram

Source: Alor, *et al.* 2014.

4.14.3.1 Define

This is the first stage of the DMAIC model. The purpose of this stage is to be clear about the behavior of the problem to be solved by the company and to define the customer's expectations for the process (Figure 3). The elements of this stage include a specific statement of the problem to be solved, describing the location and occurrence of the problematic events, as well as an initial diagnosis of the processes detailing the priority of the problem. At this stage, clearly define what is needed for a successful Six Sigma project. Defining includes identifying the customers (internal and external); identifying their needs, determining the scope of the project and the objectives of the project (Manivannan, 2007).

4.14.3.2 Measure

The Measurement stage establishes techniques for collecting data on current process performance and how well the customer's expectations are being met. At the end of this stage, there must be an information collection plan, a valid measurement system to ensure accuracy and consistency in data collection, frequency of defects and sufficient data for problem

---

[2] DMAIC methodology applied to improve existing processes

analysis (Manivannan, 2007).

4.14.3.3 Analyze

This phase of the project seeks to establish differences between the desired level of performance and quality of the process and the current level of the company's process. It is the assessment of the causes that generate a problem that requires improvement and analysis of how to eliminate or reduce these causes, this will reduce defects and process variation (Figure 3). This is done through the detailed study of the key processes of the indicator variables previously identified. The prioritization of the key processes of the indicator variables performed during the measurement phase helps to identify among the factors; the highest priority (Vootukuru, 2008).

4.14.3.4 Improve

This phase is where the project research evaluates the information from the analysis phase of the main causes of the problem and proposes various solutions to eliminate or reduce those causes (Figure 3). The solutions are analyzed and evaluated through tests, possible simulations and the design of experiments to prioritize the most optimal one. An implementation plan for the improvement is then developed with an approach that encompasses improvement management that helps the organization to accept, implement and adapt to the changes produced in the solution (Vootukuru, 2008).

4.14.3.5 Check

In this phase of the project investigation, actions are developed to support the efficient implementation of solutions during the improvement phase (Figure 3). To ensure that the improvement is effective, a strict monitoring plan that identifies the main actors involved in the process must follow. This process must be accompanied by reaction plan projects and adequate training resources to carry out the vital activities of this phase and thus meet the productivity increase based on quality and performance (Vootukuru, 2008).

4.14.3.6 Tools used in the phases of the Six Sigma Methodology

According to Domínguez (2005), the tools that can be used in the 5 phases of the Six Sigma methodology are:

Define: Affinity diagram, brainstorming, customer research, cause and effect diagram, qualitative process analysis, process flow diagram, Kano analysis, SIPOC[3] and QFD[4] (Quality function deployment).

Measure: Competency analysis, check sheets, control charts, graphing techniques, Gage R&R (repeatability and reproducibility studies), indicators, sampling, Sigma calculation, stratification and variation.

Analyze: Hypothesis testing, design of experiments, control charts, Pareto chart, regression, fault tree, reliability, simulation and process analysis.

Improve: Cost-benefit analysis, robust designs, sequential experimentation, QFD, confirmatory testing.

Control: FMEA, process control system, project closure and work planning.

5. COMMENTS

The community forest industry is a good alternative to stimulate the involvement of

---

[3] SIPOC in Spanish suppliers, inputs, processes, products, customers, etc.
[4] QFD translates as Quality Function Deployment.

communities in the better management of natural resources, while at the same time improving their living conditions, since through proper organization, administration and sustainable production, profitable results could be achieved for the company. These organizations are a source of employment that serves to mitigate the country's current economic crisis and improve competitiveness in the timber industry. Honduras is a country with abundant natural resources, a situation that provides adequate conditions for obtaining raw material for the production of sawn timber. However, industries need to make production processes more efficient in order to generate better yields, reduce costs and improve product quality to achieve better customer satisfaction and competitiveness in domestic and international markets. One of the most important aspects today is to involve total quality management in companies through continuous improvement, by means of strict processes of diagnosis, analysis, application of improvements and evaluation of production processes.

Quality management would then be the improvement of the quality of the processes, thus generating a better productivity that would lead to the positioning of a company in the markets, which would generate better employment opportunities and profits for the company.

CHAPTER 2

1. PROBLEM DEFINITION

At present, the demand is based on higher demands for quality products and accessible prices. Another fundamental aspect is the choice of customers to purchase products made of wood pellets or any other material, due to its low cost and better final finish. The most fundamental current challenge for primary sawmilling companies is to become more competitive in the market.

El Palisal Agroforestry Cooperative is a community enterprise with 13 years of experience and deficiencies in quality and yield in its production process. One of the company's objectives is to generate income for its members through the profits generated from product sales. To date, this has not been fully achieved because the company has not been able to consolidate its production process, as it has several problems such as low product quality and inadequate wood classification. This has led to rejection and a decrease in the price of the product by the company's potential clients. Another problem encountered is low yields, as of June 2015 the company only produced 210 board feet per cubic meter of roundwood, a monthly production of 21,272 board feet of lumber, which indicates that the company is below the adequate yields, demands and qualities suggested by clients.

As a result of this current situation of the company, this study focused on the evaluation of the yield and quality of wood, before and after the implementation of continuous improvements in the process of primary sawmilling of wood, being an alternative to provide technical and innovative information that leads to the generation of better productivity and provide a quality service.

2. HYPOTHESIS

By implementing the Six Sigma methodology in the Agroforestry Cooperative "El Palisal", it will be possible to improve the yield and quality of the sawn timber in the primary sawmill process, strengthening the competitiveness and generating opportunities for better profits for the Cooperative.

3. METHODOLOGY

3.1 Study area

The municipality of Yamaranguila has a total area of 31,256 ha, of which 9,818.6 hectares are under a forest management plan registered in ICF file number BE-L2-002-98-III, representing 31.4% of the municipality's area. The vast majority of these lands are under ejido ownership (approximately 90%) and a small percentage are privately owned (Ulloa, 2001).

Yamaranguila has 14 villages and 75 hamlets. According to the last official census of the National Statistics Institute (INE) for 2001, its population was 15,723 inhabitants, mostly people between the ages of 0 and 25 (INE, 2001). Yamaranguila is located at an altitude of approximately 1800 meters above sea level, with a temperature between 18^0 C to 24^0 C, which is why it has several types of forests, including cloud forest, tropical dry forest and others (Alcaldía municipal, 2010).

The research was carried out in the El Palisal Agroforestry Cooperative, located in the community of El Obispo in the municipality of Yamaranguila; specifically in the area of primary wood sawmilling. This community enterprise has 43 members, and its main activity is pine roundwood extraction and lumber marketing. The secondary product is sticks that are used by clients to make broomsticks and stakes or stakes to support some vegetable plants, as well as sawn lumber to manufacture furniture for the local market.

3.1.1 Study scenario

The legal constitution of Cooperativa Agroforestal El Palisal Limitada was framed on March 21, 2001 and on October 24, 2002, it obtained its legal status. El Palisal Cooperative is the merger of 7 agroforestry groups in the municipality and was created through agreement 1941 DE/RNC issued by the Honduran Institute of Cooperatives (IHDECOOP), with legal status number 1941, volume VII, book III of the National Registry of Cooperatives under IHDECOOP on January 3, 2002 (Salazar, 2011).

The municipality of Yamaranguila granted 9,818.5 ha of pine forest to be managed, focusing on community forestry activities. In the beginning there were some contracts for timber sales. However, the experiences were not very satisfactory, and the product was not of excellent quality due to poor training of operators and lack of knowledge of quality controls (Salazar, 2011).

3.1.2 Production history of the El Palisal cooperative

From 2011 to 2014, Cooperativa El Palisal has not been able to harvest 100% of the volume authorized by the ICF (Figure 4), due to poor planning, low efficiency and limited installed capacity of the company. These volumes were approved through operational plans under the forest management plan with ICF file number BE-L2-002-98-III.

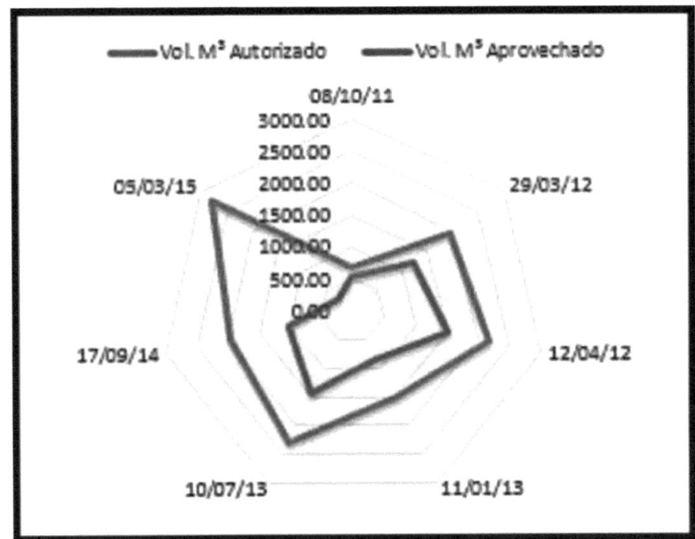

Comparison of the volume harvested by the Palisal Cooperative with the volume authorized by ICF.

Source: Yamaranguila Municipal Mayor's Office, 2015.

3.2 Equipment and tools

J Equipment: Camera, computer, datashop, stopwatch, moisture meter.

J Tools: Forceps, tape measure, meter.

J Aids: Boards, field forms, documents, hemp to measure wood curvatures, pencils and markers.

3.3 Research methodology procedure

The research work was carried out through the implementation of the Six Sigma methodology based on a 5-step DMAMC procedure (Figure 5) (Pries, 2006).

J Define the problems and situations to be improved

J Measuring indicators in the process to obtain information

J Analyze the information collected

J Improve processes according to the prioritization of problems.

J Monitor improved processes to generate a cycle of continuous improvement.

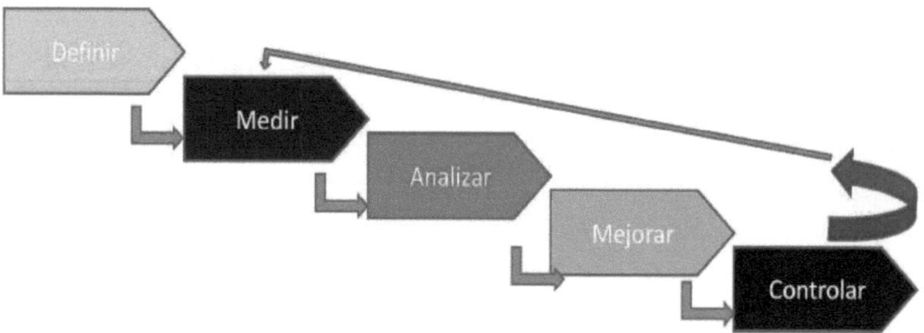

Figure 5. Six Sigma Methodology "DMAMC".

Source: Pries, 2006.

Within each stage of the methodology, tools were implemented that were used to achieve the proposed objectives, among them are: baseline, process flow, cause and effect diagram, Pareto diagram and Six Sigma level calculation with the use of the statistical program Minitab. It is necessary to mention that the cycle of application of the Six Sigma methodology was carried out first to evaluate the current process of the company and then it was applied again to evaluate the process already with the implementation of continuous improvements.

3.4 Field work

3.4.1 Define

At this stage, the problem to be improved was identified, as well as the factors that served to prioritize the problem to be addressed, the current customers and the current status of the cooperative's process. In order to carry out the characterization of the company's process, the following tools were implemented:

3.4.1.1 Definition of a baseline of the company's production process.

It was carried out according to existing documentation in the cooperative taking into account data from 2014 to June 2015, the documents that were used as a source of information were; business plan, forest management plan, annual operating plan and monthly inventories of production and sale of wood, in addition to the qualitative and quantitative data collection of the primary sawmilling process and wood in storage, to improve the information a meeting was held with the plant manager and the sales manager in order to acquire relevant information on volumes produced, sales volumes, demands and customer requirements.

3.4.1.2 Definition of the current process flow diagram

The primary sawmilling process was defined in a participatory manner by means of a simple graphic representation of the basic phases and operations that constitute it, identifying its inputs, processes and outputs of the company.

In order to know the current status of the primary sawmill process of the cooperative, the methodology to be followed was explained to each of the operators and members of the board of directors, then for reasons of time and to avoid delays in daily production, each of the operators of the yard area, sawmill process, storage, classification and drying of wood were interviewed individually, with the purpose that each person described the activity performed during the process, noting the type of machine used, measurements, times and method used in the production system.

3.4.1.3 Cause and effect diagram definition

The cause and effect diagram was made in conjunction with the machinery operators and employees of the cooperative, in addition to which key people with experience in timber industries of the projects that currently support the cooperative were invited, including Catie-Finnfor, Mosef and an independent consultant, in order to define the causes that affect the production process of sawn timber and consequently the reason for the occurrence.

The following steps were followed to achieve the objective:

Step 1: The problem identified (low yield and poor timber quality) was explained in detail.

Step 2: Participants were trained on the use of the 6M diagram consisting of: Measurement, Material, Labor, Environment, Methods and Machines. Then it was applied to each area of the process sequentially (yard, sawmill process, storage, classification and drying of wood), in order to be clear about the causes and effects in each of the areas according to the categories that were appropriate to it.

Step 3: Using the "brainstorming" technique, each participant gave their opinion according to their experience and knowledge, in such a way that each of the causes affecting the process were defined.

Step 4: To define the prioritization of the causes found, each of the causes found was rated with a score from 1 to 10, taking into account that 1 is not a priority, 10 is a high priority, thus getting a clearer idea of the causes that should be attacked first.

Step 5: Once the causes that generate the problem were defined, they were ordered in a format according to category and qualification.

3.4.2   Measure

Based on the data obtained in the definition stage and according to the most common causes that affect the primary sawmilling process, methods were applied to collect data on the current performance of the sawmilling process:

3.4.2.1 Time and motion study

The company's current process flow was taken as a reference, then we proceeded to define the times in minutes that each of the activities of the company's production process entails, for which the following steps were followed:

Step 1: Determination of the sample

In order to carry out the time and motion study, the number of observations or cycles that had to be made before being able to determine the standard time for each operation was calculated. For the purposes of the study, each production cycle consisted of one work shift; the company is currently processing approximately 5 cubic meters per shift. To calculate the standard deviation, the amount of wood processed in board feet in the last 26 days (production from May 15 to June 15) was taken into account, and then a table of frequencies and the respective calculation of the number of samples was made (Equation 1).

Equation 1. *Sample size calculation (cycles)*

$$n = \frac{4\left(t_{0.95}\right)^2 s^2}{I^2}$$

Where: Sample size (n), Student's t-distribution value at 95% confidence (t0.95), standard deviation (S), 5% sampling error (I).

Step 2: After defining the sample size, we proceeded to take the times of each of the operations of the process for which a field format was used (Annex 4.4), with the purpose of defining the following: Total time, average time, qualification factor, normal time and average time of the operation: Total time, average time, qualification factor, normal time and average time of the operation. To determine the qualification factor, the Westinghouse System was used, which relates the operator to four factors such as: ability, effort, conditions and consistency (De La Roca, 1994).

Step 3: Once the times of each operation of the process were obtained, the standard time was calculated, which was used to define the overall time of the operation (Equation 2).

Equation 2. *Calculation of standard time*

Standard time = TN/ (1-Supplementary time in %)

Where

TN= Normal Time (multiplication of the qualification factor by the average of the times). Rating factor according to the Westinghouse System used was 1.3 and 1.2 for the sawmill activity, truck unloading, machine check and wood storage, the variation is due to the different people in charge.

Supplementary time= understood as the percentage assigned by the evaluator considering the needs of people, fatigue and special times in the process (Moori, *sf*).

3.4.2.2 Quality sampling

Quality sampling was carried out with the objective of defining the number and percentage of defects in each board, for which each defect was quantified by determining the damaged portion in percentage and the number of defects that the piece of wood may have (Annex 4.6). For the quality sampling, a sample was taken, which was calculated using the number of pieces produced in the last 26 days (production from May 15 to June 15), and then a frequency table and the respective calculation of the sample were made (Equation 1). In this case the number of defects was used to calculate the Six Sigma level and the percentage to define the quality grades of the wood. The quality sampling was carried out in the two scenarios proposed before and after the application of the process improvement.

3.4.3 Implementation of improvement measures

Taking into account the data collected and analyzed in the definition, measurement and analysis stage, the improvement activities were prioritized together with the members of the board of directors, applying criteria such as: ease of carrying out the experiment, space, time and cost-benefit of investment. After reaching a mutual agreement, we proceeded to the practical application of a combined improvement, which contemplated 4 sub-improvements, among them: classification of round wood in bacadia, training in cutting diagramming, preventive maintenance of the saw and change of damaged parts of the saw and conditioning of the drying area, after the improvement was applied, it was re-evaluated in terms of quality and performance two months later, following the phases of: measurement and analysis. Once the data were obtained before and after the application of the improvement, a comparison of the process was made to define the differences that could be obtained.

3.4.4 Control

During this stage, a process control strategy was developed, based on the results of the four previous stages, so that the improvement applied to the process could be implemented efficiently. The control was carried out during the adaptation time of the improvement, for which a monthly plan was created, indicating the processes and activities, which were supervised by a member of the board of directors.

3.5 Office work

3.5.1 Coordination and management work

Four meetings were held: During the first one, the research was socialized before the members of the El Palisal Cooperative. In the second, the cause and effect analysis was defined. In the third meeting, the results of the process analysis, the selection of the improvements to be implemented and the organization of the control work after the implementation of the improvement measures were presented. The last meeting was held to present the results of the investigation.

3.5.2 Situation analysis of the cooperative

3.5.2.1 Preparation of Pareto charts

Based on the data found in the development of the cause and effect analysis, and taking into account the categories found by area with the application of the 6M technique, the analysis of the company was carried out. Once the data on causes by area were available, the Pareto charts were constructed (Figure 6) according to the following steps:

a. Data sorting

The data were tabulated and arranged in a frequency table in the Minitab statistical program.

b. Calculation of percentages

The relative percentage of each cause was calculated with respect to the total frequencies found (Equation 3). This study is referred to as percentage weights.

Equation 3. *Calculation of relative percentage*

$$\text{Porcentaje relativo} = \frac{\text{Calificación de la causa}}{\text{Total de calificaciones}} \times 100$$

c. Calculation of accumulated percentages

The percentages of each factor were calculated by adding them consecutively.

With this information, the percentage of times that the problem represents and would be eliminated if effective actions were taken to eliminate the main causes of the problem.

d. Identification of the areas to be plotted

- **Identification of the axes**

On the horizontal axis, the factors or categories were noted from left to right, in decreasing order of their rating. The left vertical axis is graded to show the number of data observed (the rating of each cause), and the right vertical axis shows the cumulative relative percentage.

- **Representation of bars**

The bars or rectangles corresponding to the different causes were plotted according to the rating.

- **Percentage representation**

The points representing the cumulative relative percentage were placed, taking into account the graduation of the right vertical bar; the points were placed starting from the origin and then at the position corresponding to the right end of each bar and a curve was drawn linking these points (Figure 6). In this way the relative percentage curve is plotted, it is necessary to emphasize that in this example several categories are plotted, a situation that varies in the analysis of the research data according to the productive area.

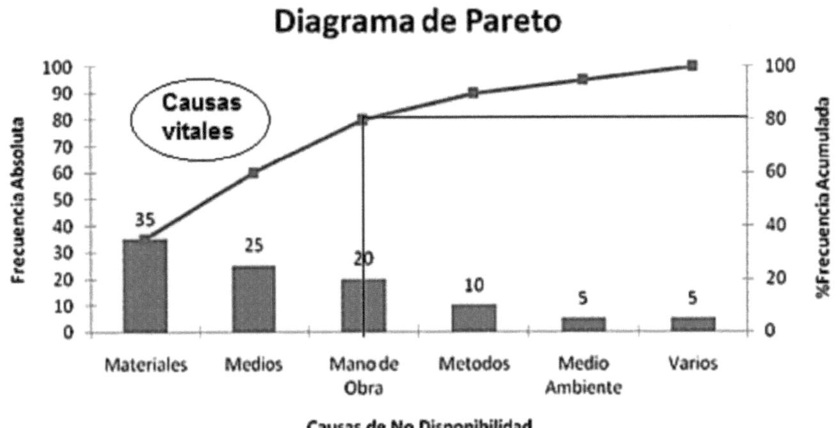

Figure 6. Pareto chart.

e. Pareto analysis

After graphing the categories based on the frequency of the causes found, the analysis of the main causes and their prioritization was carried out. Based on the Pareto diagram, which considers that 20% of the causes produce 80% of the effects or failures, decisions were made for the selection of the causes to intervene.

3.5.2.2 Elaboration of the process diagram.

The process diagram was made using the standard set of elements and symbols (Figure 7) defined by the American Society of Mechanical Engineers cited by Abraham (*n.d.*). These are described below:

SYMBOLOG1A	MEANING	DESCRIPTION
	Operation	Activity that involves a physical or chemical transformation into a product or component.
	Delay	Time period in which a component of the product is waiting for some operation, revision or transfer.
	Storage	To store a product or input in the warehouse until it is needed for use or sale.
	Inspection	It is to make comparisons or verifications of characteristics against quality and quantity standards.
	Transportation	Any movement that is not part of an operation or inspection.
	Combined operation	When an operation and an inspection are carried out simultaneously.

Figure 7. Symbology to define the process diagram.

Source: Abraham, *n.d.*

Using the data from the time and motion study, a process diagram, also known as a route diagram, was drawn up. This provided information to better understand the process in terms of distances and times, in order to propose improvement and control mechanisms.

3.5.3   Improvement selection

The brainstorming technique was applied with members of the board of directors and operators, based on the causes found and analyzed, we proceeded to make an action plan, which will include those activities that lead to the correction of the problem, it is necessary to emphasize that in this activity we prioritized those activities achievable in ease of experimentation, space, time and cost-benefit of investment.

3.5.4   Comparison of lumber yield and quality before and after the application of continuous improvements in primary wood processing.

3.5.4.1   Sigma level calculation of the process for two scenarios

The Sigma level calculation was used so that the implementation of the improvement could be controlled and carried out in a better way, so that it would be sustainable over time. This calculation was made in the measurement and control stage, for which the following information was taken:

- **Default opportunities (OD)**

The number of procedures and operations in which errors are made was quantified, understanding that an error is a defect that results in the rejection of the sawn timber.

- **Units processed (UP)**

It was taken as the sum of all the processed board feet that were evaluated in the quality sampling of dimension lumber production. That is, the number of board feet evaluated.

- **Defects (D)**

It was the total number of defects found in the quality sampling of dimensioned lumber processed during the stipulated production cycles.

- **Number of defects per opportunity (DPO)**

Total defects were divided by the number of defect opportunities per units produced and evaluated in the sample (Equation 4).

Equation 4. *Calculation of the number of defects by opportunities*

$$DPO = \frac{\sum D}{\sum OD \times \sum UP}$$

Where: Defects per opportunity (DPO), Defects (D), Defect opportunities (OD) and Units processed (UP).

- **Performance (R)**

It indicated the probability that the wood is free of faults or defects (Equation 5).

Equation 5. *Calculation of yield*

R=1-DPO

Where: (DPO) defects by opportunity

- **Process sigma (a)**

The Sigma level defines the percentage of quality in the processes carried out by the Palisal cooperative, in other words, it is the current state of performance of the process with respect to its variability (Equation 6).

Equation 6. *Process Sigma Calculation*

$\sigma =$ Inverse standard normal inverse distribution of yield + 1.5

3.5.4.2 Statistical analysis

The completely randomized design (CRD) was used for the research, because it is a design that is adapted when working in homogeneous experimental units (EU), these experimental units have the same probability of receiving any treatment t, another aspect is that it is applicable to studies with a low number of treatments.

Mathematical model used:

$$Y_{ij} = \mu + \tau_i + \varepsilon_{ij} \qquad i: 1,2,3,\ldots,t \; y \; j: 1,2,3,\ldots,r$$

Where:

Y_{ij} : ij-th observation of treatment i and experimental unit j.

μ: Overall average.

τ_i: Effect of the i-th treatment.

ε_{ij}: Experimental error.

Experimental design approach:

UE= Primary wood sawmill plant.

Treatments:

Control= Process without enhancement, Treatment 1= Combined enhancement

ANDEVA table

Source of variation	G.L
Treatments (t)	t-1
Error (E)	t (r-1)
Total	t*r-1
t= 2, r= 10 cycles production	

Due to the variability of the extracted data, variance analysis was performed in order to find significant differences in the two scenarios (Scenario 1 process without improvement,

Scenario 2 process with improvement). The analysis of variance was used because it is easy to perform and the results are easy to interpret. Similarly, for the analysis of wood defects, proportion tests were carried out so that significant differences could be observed in the two scenarios. Another input that facilitated the statistical analysis was the use of process control charts.

3.5.5   Methodological design of continuous improvement

The methodological design was built taking into account the priority of each one of the causes, that is to say, based on the priority of the problem and the economic capacity of the Agroforestry Cooperative El Palisal, this methodological scheme is an input for the members of the board of directors to make decisions from a technical point of view, as well as for those non-governmental organizations interested in supporting the growth of the community enterprise.

CHAPTER 3

RESULTS

1.1 Baseline of the Palisal cooperative focused on production and marketing.

El Palisal Agroforestry Cooperative is organized by a board of directors presided over by 15 members and a general assembly composed of 43 active members from 11 communities in the municipality of Yamaranguila Intibucá. The company currently provides various benefits to its members, including employment, credit, incentives, and housing support.

In the production and administration processes, the cooperative generates 39 permanent jobs involved in the entire production chain. As of June 2015, the company processed an average of 101 cubic meters of roundwood per month, producing 21,272 board feet per month, according to these figures the company manages yields of less than 210 pt/m^3 . With this production and yields obtained to date, the company has not been able to meet the demand, which oscillates above 27,000 board feet per month. The company's lumber buyers include Maderera Espinoza, COHAPIL, HONDULOG, SERPROMA, various buyers in the region, and carpenters near the company, among others. The most common customer requirements are: wood without blue stain, without excess knots, without rotting, without gems, wood without resin pockets, wood that is not bowed or cracked, not chewed by insects, among others, yet the cooperative has not been able to offer a product that meets all of the customer's requirements. To date, the company has not had any returns due to wood defects, but there has been a decrease in the cost of the product, varying between 1 and 2 lempiras per board foot (Lps/pt), a situation that has become critical since the cost per board foot has remained between 9 and 9.5 lempiras and a sales price of 11 lempiras per board foot. These problems are the result of the company's lack of classification processes, continuous training, quality controls, and poor machinery and infrastructure.

1.2 Results of meetings with personnel working in the lumber production process.

1.2.1 Lumber production diagram

The diagram was made with the participation of the personnel involved in the lumber production process of the cooperative. In this phase, we sought to have the process flow of the company without knowing in depth the causes and effects that generate the low quality and yield of the lumber. The objective of the process diagram is mainly to understand the production system and begin to identify potential flaws in the process (Figure 8).

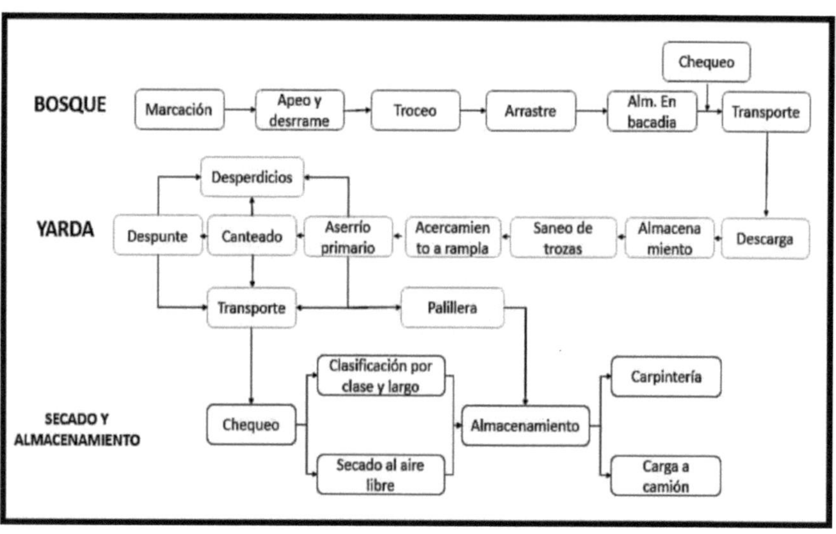

Figure 8 Diagram of lumber production at El Palisal cooperative.

8.2.2 Cause and effect diagram for the forest

The application of the 6M helped to identify and classify the causes for each stage of the lumber production process. In the forestry area, it was identified that the biggest problems are related to poor personnel training. The causes that most affect lumber quality and yield are: lack of experience in roundwood grading, lack of quality control, volume production without grading, and lack of knowledge of the measures suggested by the market (Table 1 and Annex 4.1).

Table 1. *Summary of the main problems identified in the forest.*

Components Qualification Causes	Measurements	Labor	Method	Machine
Causes	Lack of knowledge of measures suggested by the market	Little experience in roundwood sorting	Log stacking in blue whiting	Low number of chainsaws in operation
Rating	8	10	4	4
Causes			Use of animal traction for hauling	
Rating			3	
Causes			Volume production sort	
Rating			9	
Causes			No quality control in forest	
Rating			10	
Total qualification	**8**	**10**	**26**	**4**

Note: The Material and Environment components are not reflected because they are not within the area.

8.2.3   Cause and effect diagram yard

With the application of the 6M methodology, it was possible to identify that the primary sawmilling area is where the greatest number of causes that affect the quality and yield of sawn timber are generated, compared to the forest, storage and drying areas. The causes are related to deficiencies in the machine, methods used and the material used as raw material. The causes that have the greatest impact include: defective logs, lack of experience in cutting diagrams, lack of spare parts (saw parts with wear), lack of preventive maintenance of the main saw, and lack of classification of roundwood as sawnwood (Table 2 and Annex 4.2).

Table 2. *Summary of the main problems identified in yarding and sawmilling.*

Components Qualification Causes	Material	Labor	Environment	Method	Machine
Causes	Logs with physiological defects and anthropogenic	Little experience in cutting diagramming	Yard planting with unevenness	No preventive maintenance plan for the sawmill	Saw blades with wear
Rating	10	9	4	9	10
Causes	Logs over dimensioned to the capacity of the saw.	Low experience in truck unloading	Little space for internal mobilization	Nohay round wood grading	Saw head bearings with wear
Rating	7	4	3	8	8
Causes	Logs with incorrect dimensions			Nohay wood sawing planning	Sheaves with unevenness and wear
Rating	6			8	8
Causes	Too long in storage round wood			Do not check the mobilization	Reverse unleveling of sleepers
Rating	5			2	7
Causes	Logs stored directly on the ground			Logs stored in disarray	Nohay machinery for edging and trimming
Rating	5			6	6
Total qualification	**33**	**13**	**7**	**33**	**39**

Note: The measurement component is not denoted because no measurements are made in the area.

8.2.4   Cause and effect diagram drying and storage

With the application of the 6M methodology in the drying and storage area, it was possible to identify that most of the causes are related to the methods used, among the most important we can mention: very high pallets generating curvatures in the first stored boards, wood stored less than one foot above the ground, most of the wood does not have a drying process and there is no experience in drying and preservation methods (Table 3 and Annex 4.3). Although the methods used generate the greatest number of important causes, it is necessary to mention that the high moisture content in the wood generates the problem of blue stain in a significant proportion of the sawn wood.

Table 3. *Summary of the main problems identified in drying and storage.*

Components Qualification Causes	**Material**	**Labor**	**Environment**	**Method**
Causes	Wood is not graded by quality	Little experience in wood classification	Presence of moisture in the soil	Drying of wood on easel
Rating	4	4	7	6
Causes	High moisture content in wood	Little experience in timber storage	Poor ventilation in the storage area	Very tall pallets exceeding 4 feet in height
Rating	8	7	3	9
Causes		Little experience in drying and wood preservation	Little space for circulation	Maderaalmacenada directly from the sawmill area
Rating		8	3	7
Causes				Few spacers in long wood
Rating				2

Causes				Lumber stored less than 1 foot off the ground
Rating				7
Total qualification	12	19	13	31

Note: The measurement component is not denoted because no measurements are made in the area.

8.2.5 Pareto diagram.

The graphs are based on the so-called Pareto principle, known as the "80-20 Law" or "vital causes, insignificant causes", which defines that only a few causes (20%) generate most of the problem (80%); the rest form a minimal part of the total problem (Gutiérrez, 2010). The data obtained from the application of the 6M technique, used for cause and effect analysis, were used to prepare the Pareto charts.

8.2.5.1 Pareto analysis for the forest area

According to the information shown in the Pareto chart for the forest area (Figure 9), there are two causes that should be of importance when making decisions: It is necessary to train forest personnel in roundwood grading methods and give clear instructions to the hill manager or the person in charge of the operational plan administration to carry out quality control by supervising and providing instructions to the other operators, so that roundwood with a high degree of defects can be avoided. It is important to take into account that the quality of the raw material acquired from the forest will determine the quality of the sawn timber and the yield per cubic meter. Another advantage of producing a better quality of raw material in the forest is that production costs can be reduced and therefore profits can be improved.

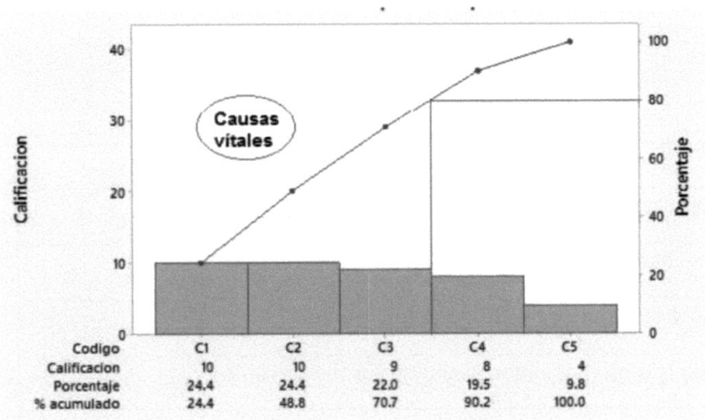

Codigo	C1	C2	C3	C4	C5
Calificacion	10	10	9	8	4
Porcentaje	24.4	24.4	22.0	19.5	9.8
% acumulado	24.4	48.8	70.7	90.2	100.0

Figure 9 Pareto plot for the forest harvesting area.

Where:

CodeDetail

C1 Little experience in roundwood sorting
C2 No quality control in forest
C3 Unclassified volume production
C4Unfamiliarity with   measures suggested by the market
    C5Others

1.2.5.2   Pareto analysis for yard area

According to the analysis, the most important causes for the yarding and primary sawmilling area are: logs with a high percentage of defects, saw blades with wear, little training in cutting diagrams and no preventive maintenance plan for the main saw (Figure 10).

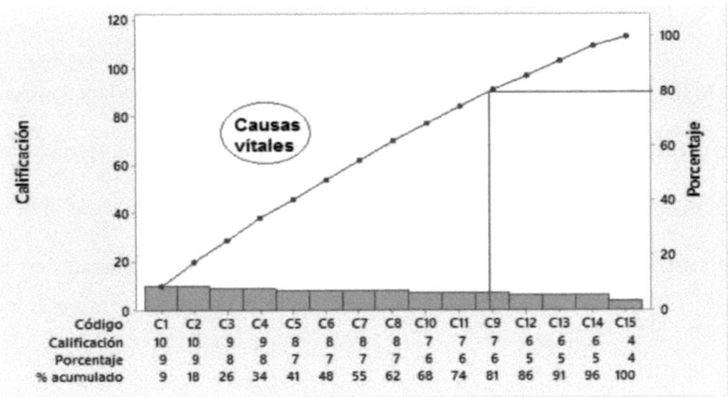

Código	C1	C2	C3	C4	C5	C6	C7	C8	C10	C11	C9	C12	C13	C14	C15
Calificación	10	10	9	9	8	8	8	8	7	7	7	6	6	6	4
Porcentaje	9	9	8	8	7	7	7	7	6	6	6	5	5	5	4
% acumulado	9	18	26	34	41	48	55	62	68	74	81	86	91	96	100

Figure 10 Pareto chart for yard area

Where:

Code	Detail
C1	Logs with physiological and anthropogenic defects
C2	Saw blades with wear
C3	Little experience in cutting diagramming
C4	No preventive maintenance plan for the sawmill
C5	No roundwood grading
C6	No planning in the sawing of timber
C7	Saw head bearings with wear
C8	Sheaves with unevenness and wear
C9	Logs over dimensioned to saw capacity
C10	Reverse unleveling of sleepers
C11	Chassis with a maximum length of 20 feet
C12	Logs with incorrect dimensions
C13	Logs stored in disarray
C14	There is no machinery for edging and trimming.
C15	Others

1.2.5.3 Pareto analysis for the drying and storage area

According to the analysis in the drying and storage area, the most important causes are: high pallets that exceed four feet in height, high moisture content in the wood storage area, and lack of experience in drying and preserving wood (Figure 11). It is important to mention that the infrastructure conditions in this area are not adequate, which generates greater exposure to wood defects.

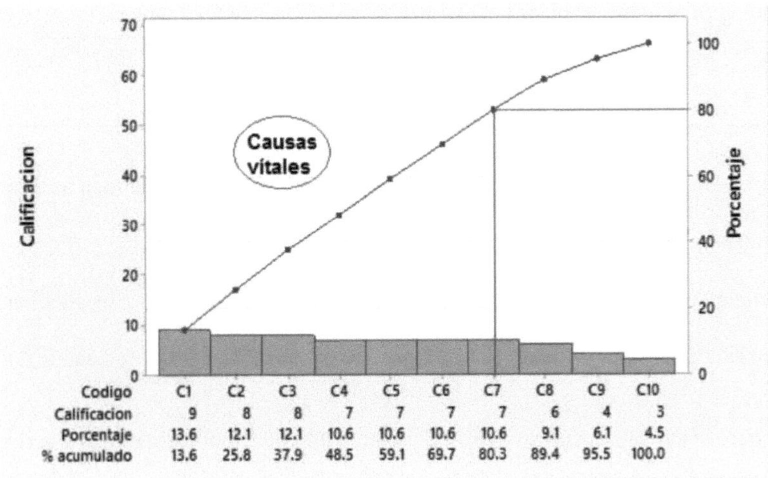

Figure 11 Pareto chart for drying and storage area

Where:

CodeDetail

C1 Very tall pallets exceed 4 ft. in height
C2 High moisture content in wood
C3 Little experience in drying and preserving wood
C4 Little experience in timber storage
C5 Presence of soil moisture
C6 Lumber stored directly from the sawmill area
C7 Lumber stored less than 1 foot above the ground
C8 Drying of wood on trestle
C9 Timber is not graded by grade
C10 Other

1.2.6   Selection, implementation and control of process improvements

In a participatory manner with members of the board of directors, based on the analysis of quality and performance, an improvement was selected, implemented and controlled. For the purposes of the experiment, a combined improvement involving the three work areas (forest-yard-drying and storage) was chosen (Annex 4.23). This improvement was prioritized according to the need found, costs, space and time for better applicability and control. The process improvement implemented is detailed as follows:

a. Grading of roundwood placed in bacadia, to avoid the arrival of wood with defects at the sawmill.

b. To train operators on the subject of roundwood cutting diagrams.

c. Improve the conditions of the drying area (roofing) to avoid defects such as: blue stain, wood cracking and others.

d. Training in the use and preventive maintenance of the main saw, in addition to the replacement of damaged parts of the saw (saw blades, brake flywheel, head bearings).

The amount of the improvements as a whole was 30,100.00 lempiras, taking into account that the company generates a profit of 1.5 lempiras per board foot produced, the recovery period of the investment is one month.

The training process was carried out to improve the skills of the operators, although the evaluation period is relatively short. According to Obregón, *et al.* (2008), the training evaluation process should be carried out one year after the training has been applied, but intermediate evaluations can also be carried out at established periods of time.

For the control of the improvement implemented, a plant supervisor and a forest supervisor (hill manager), both members of the board of directors, were established to control the process and the proper application of the improvement.

With the combined improvement, we sought to attack those vital causes that generate a greater potential in the problem, taking into account some of the causes that are within the 20% and generate a greater percentage of the problem, among them are mentioned:

- Little experience in roundwood sorting
- No quality control in the forest

- Logs with defects

- Saw blades with wear

- Little experience in cutting diagrams

- No roundwood sorting is performed

- Sheaves and bearings with wear

- Very tall stowage exceeding 4 feet in height

- High moisture content in wood

- Little experience in wood storage

1.3 Time and motion study

1.3.1   Sample size calculation

Using the daily production data in board feet between May and June (26 production days), the number of board feet produced per day was calculated, resulting in 680.12 board feet, the standard deviation was 159.52 board feet, the t value for 25 degrees of freedom and a significance level of 0.05 is 2.06 and for an error of 5%, the value of the sampling error is 9.47. With these values the sample size was:

$$n = \frac{4(2.06)^2(159.52)^2}{(9.47)^2} = 4816.42/680.115 = 7.08 \approx 8$$

The sample size calculation based on the measurements made was 8 cycles. Even so, ten measurements per cycle were considered to achieve better precision.

1.3.2   Calculation of standard times for each operation before and after the application of the process improvement.

The objective of measuring the work or the measurement of times and movements was to define strategies or visualize problems in the process flow, in order to increase the efficiency of the process, eliminate unproductive times, determine the workload for the operators, among other aspects. Moori, (*n.d*) defines standard time as the time in which any task can be carried out by a trained person, developing a normal activity according to the established methodology and including the tolerances related to delays that are out of the worker's control.

Table 4. *Standard process times before improvement application.*

Variables	Elements				TOTAL
	I II		III	IV	Minutes
Average operating time	271.3	33.8	46.8	107.8	459.7
Normal time	341.9	41.5	58.1	130.5	571.9
Standard time	468.3	56.9	79.5	178.7	783.5

Where:

	Primary sawmilling
	Truck unloading
III	Machine stops
IV	Wood storage

In the normal process of the Palisal cooperative, there are high machine downtimes due to the fact that there are usually changes of saw blades, changes of damaged bands and log sanitation. For the average activity without applying improvements in the process, the standard time is 783.5 minutes or 13.05 hours per production cycle, this includes the average operation time, evaluations[5] of the process and supplementary times[6] (Table 4).

Table 5. *Standard process times after improvement application.*

Variables	Elements			TOTAL
	I II	III	IV	Minutes

Average operating time	181.6	30.8	25.2	101.0	338.6
Normal time	228.8	37.9	31.3	122.2	420.2
Standard time	313.4	51.9	42.8	167.4	575.6

Where:

	Primary sawmilling
	Truck unloading
IIIP	Machine stops
	VATwood storage

In the primary sawmill process with the inclusion of a combined improvement, there is a difference in the standard time compared to the process without improvement. In this situation the standard time is 575.6 minutes or 9.6 hours, the greatest amount of time decreased is reflected in the sawing process and the decrease in machine downtime (Table 5).

1.3.3 Path diagram

The time and motion study resulted in a route diagram of the lumber production process, which was made using the symbols defined by the American Society of Mechanical Engineers for better interpretation (Figure 12). The distances between each sub-process are shown in the diagram.

When analyzing the diagram, it is clear that the sawn timber production process does not present a linear form of production, which would be ideal in order to save time and movements in each sub-process.

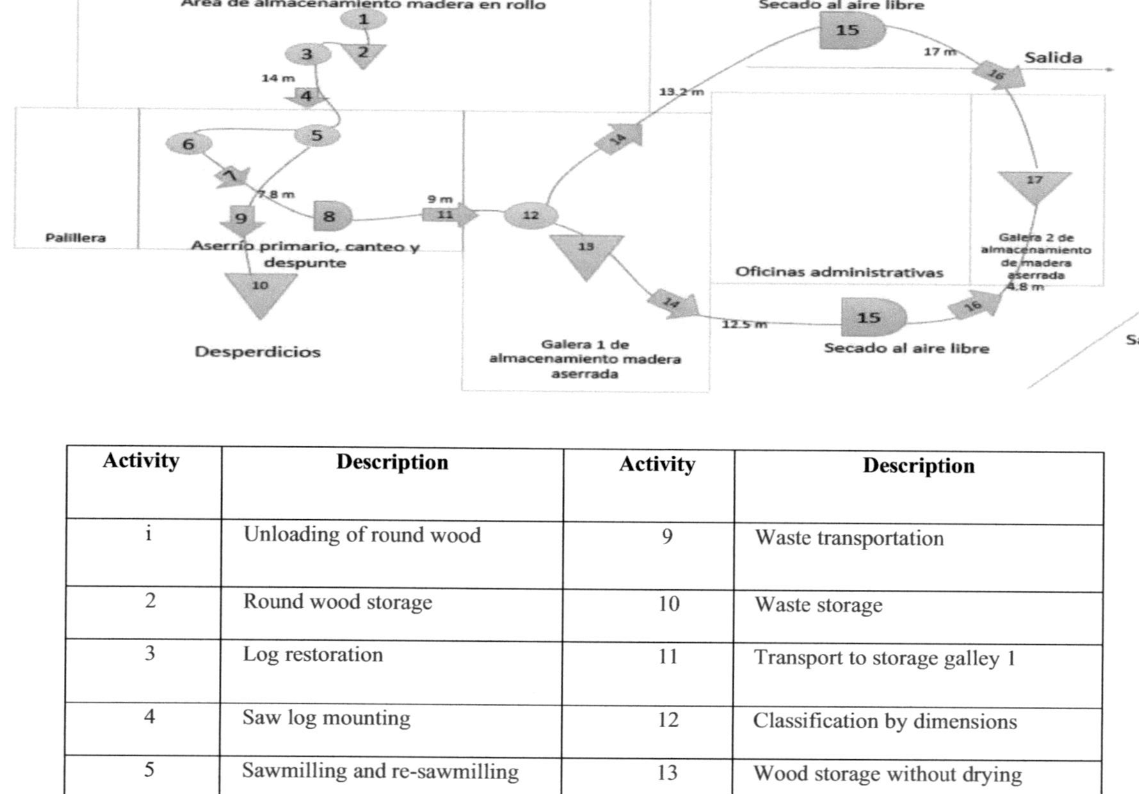

Activity	Description	Activity	Description
i	Unloading of round wood	9	Waste transportation
2	Round wood storage	10	Waste storage
3	Log restoration	11	Transport to storage galley 1
4	Saw log mounting	12	Classification by dimensions
5	Sawmilling and re-sawmilling	13	Wood storage without drying

6	Chainsaw trimming	14	Transport to drying yard
7	Timber transportation	15	Air drying
8	Initial stacking of sawn lumber	16	Transport to storage galley 2
		17	Storage of air-dried lumber

Figure 12 Route diagram for lumber production at the El Palisal cooperative.

1.4 Quality sampling

Using the daily production data in units produced between May and June (26 production days), the number of pieces produced per day was determined, resulting in 64.42 pieces, the standard deviation was 17.66 pieces, the *t value for* 25 degrees of freedom and a significance level of 0.05 is 2.06 and for a 5% error, the value of the sampling error is 11.07. With these values the sample size was:

$$n = \frac{4(2.06)^2(17.66)^2}{(11.07)^2} = 43.2 \approx 44 \text{ piezas}$$

The sample size calculation based on the measurements made was 44 pieces (88 pieces in total). Even so, 603 (469 pieces for the scenario without improvement and 134 pieces for the scenario with improvement) measurements were considered to achieve better accuracy.

1.4.1 Frequent defects encountered in the production process without improvement

According to the quality sampling, an average of 34% of the wood still has defects, from the highest to the lowest percentage: knots, resin pockets, marrow, gems and blue stain (Figure 13 and Annex 4.8).

In order to have a better idea of the quality of the wood, it was divided into two categories[7] A and B, in this case 45.6% are category A and 54.4% are category B.

---

[7] The categories were divided according to the COHDEFOR classification, where category A presents a percentage of less than 20% and category B a percentage between 20% and 100% of the wood affected.

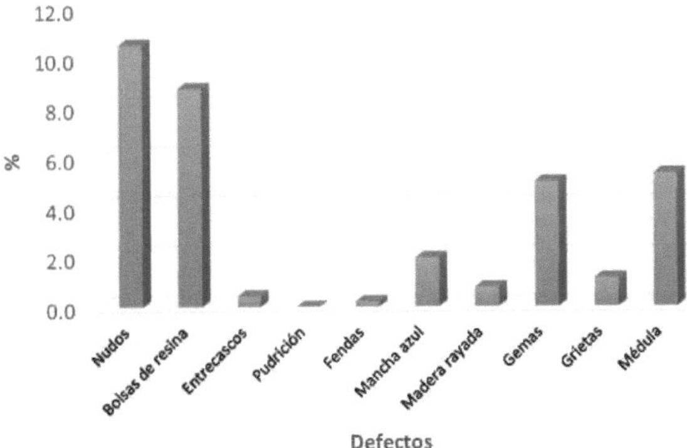

Figure 13 Percentage of defects found in the process without improvement.

1.4.2 Frequent defects encountered in the production process with improvement

According to the quality sampling, an average of 23.3% of the wood has defects, from highest to lowest percentage: knots, resin pockets, gems and marrow (Figure 14 and Annex 4.10). In terms of categories, 58.2% are category A and 41.8% are category B.

The decrease in defects can be attributed to log grading, quality control and cut layout.

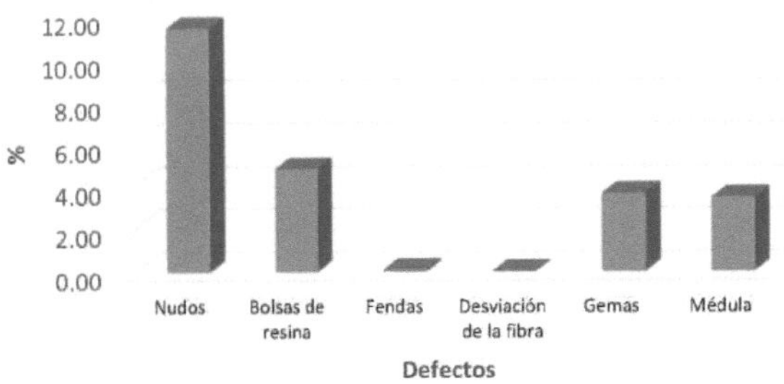

Figure 14 Percentage of defects found in the process with improvement.

14.4 3 Process capability control

A production process is subject to a number of random factors that allow it to produce products that are unequal in terms of dimensions, defects or other characteristics. This instability is not desirable in a production process and the objective should be to reduce it to a possible minimum or at least to keep it within standard limits. Statistical process control is a useful instrument to achieve the desirable objectives in a company. It can be said that this tool contributes to the improvement of the quality of the manufacture of any product. A process that maintains an acceptable variability is within the range of 3a to 6a. In other words, it maintains a capability index (Cpk) greater than 1 (Falco, 2006).

The capacity of the process without improvement of the Palisal cooperative in relation to the defects found was Cpk equal to 0.39 (Annex 4.15), indicating that the process is relatively low and that there are a number of defects capable of generating a low quality product. In the measurement of the process capability with improvement after almost two months, a capability index (Cpk) equal to 0.67 was achieved (Annex 4.16), in relation to the previous scenario the process manages to improve but the process is still low since the capability index should be greater than one.

14.5 Yield study

The Palisal cooperative according to the sampled data (10 cycles equal to the time and movement study) before the application of the improvement had an average yield of 229.7 board feet per cubic meter, this represented a percentage of 54.2% of wood used per cubic meter processed, this yield is even higher than that found in the baseline, it may be due to the influence of supervision or another aspect. After the application of the improvement combined, the average yield increased to 244.2 board feet per cubic meter, representing a percentage of 57.6% of wood utilized per cubic meter (Annex 4.14). In other words, with the application of the combined improvement it was possible to increase the yield by 14.5 board feet, representing a profit of 159.5 lempiras per cubic meter processed.

14.6 Sigma level calculation for the lumber production process pre- and post-improvement.

To calculate the Sigma level, 603 pieces of wood (boards with different dimensions) were measured according to the normal production of the company. To determine the Sigma level of the process before improvement, 469 pieces were measured and for the process with improvement, 134 pieces were measured. Within the process flow, 10 error-prone activities (OD) were defined. Based on the proportion of defects calculated with the quality samples, the units processed (UP), the number of defects (D) and thus the calculation of the Six Sigma level were estimated (Table 6).

Table 6. *Sigma of the pre- and post-improvement process.*

SIGMA OF THE PROCESS		
	Pre. improvement	**Post. improvement**
Default opportunities	10	10
Units processed	469	134
Defects	1380	383
DPO	0.294	0.286
Process performance	70.5%	71.4%
DPMO	294,243	285,821
Process sigma	2.4	2.7

This means that the process for lumber production at Cooperativa El Palisal maintains a distance of 2.7 standard deviations between the process average after the application of the improvement, in other words, according to process performance, 71.4% of the lumber produced maintains adequate quality, but even so, the process is not within acceptable parameters, which vary between 3 and 6 standard deviations. Compared to the results obtained in the process without improvement, the data show a tendency to improve from 2.4 to 2.7 standard deviations, so it is necessary to continue monitoring and controlling the improvements implemented.

14.7 Methodological design of continuous improvements for the primary sawmill process.

The methodological design is based on the cooperative's investment possibilities, intervention logic to improve process capacity and yields, as well as an investment, implementation and control analysis.

The methodological design proposes improving processes from the beginning of the production chain, taking into account that the yield and quality of the products depend on the quality of the raw material acquired. For the same reason it is necessary to start quality management from the forest to the storage of the finished product (Figure 15). It is necessary to mention that the costs of implementing continuous improvements may vary according to the time of application, personnel or institution hired.

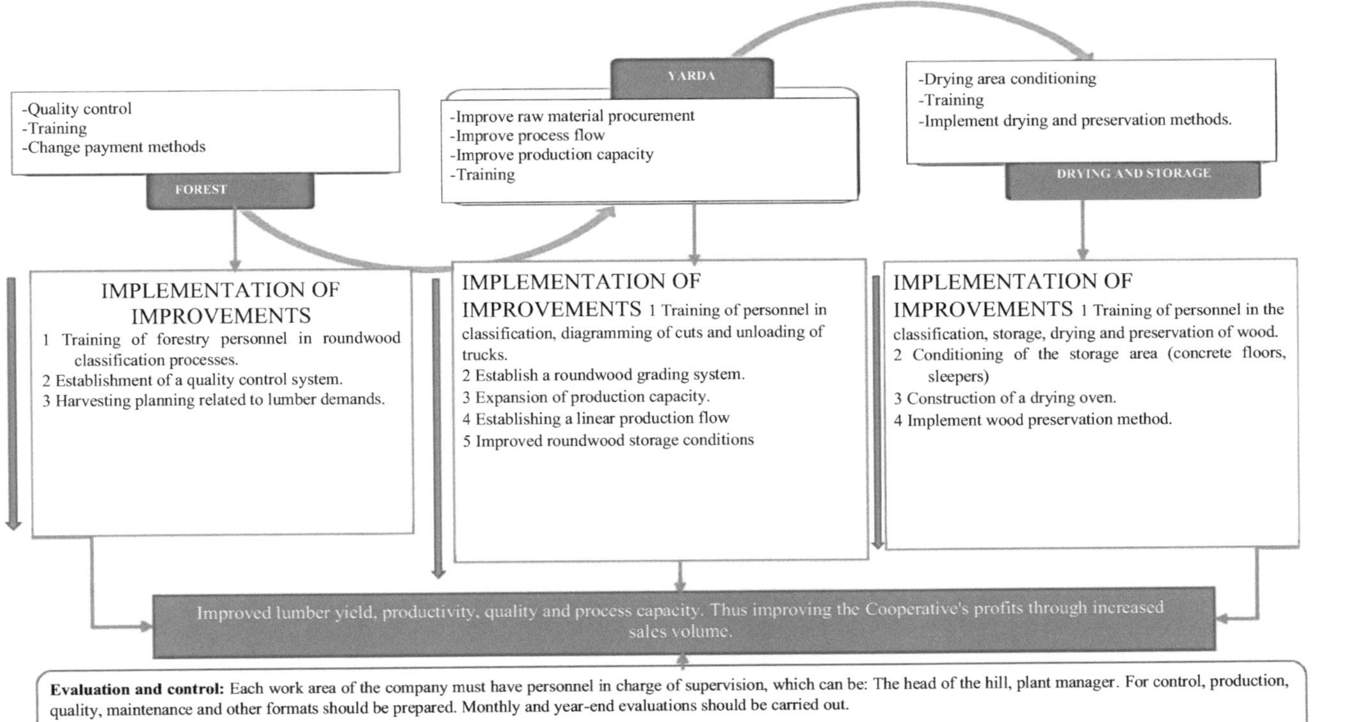

Figure 15 Methodological design of continuous improvement

2. DISCUSSION

According to the analysis of causes and effects, El Palisal agroforestry cooperative's main problems are related to the following: raw material used, machinery installed and methods used. These aspects are related to personnel training and investment possibilities. In addition, the company does not have a linear production process, a situation that generates greater investment of time and cost in production processes.

With the implementation of the combined improvement, which included 4 sub-improvements, among them: classification of round wood in bacadia, training in cutting diagramming, preventive maintenance of the saw and replacement of damaged parts of the saw and conditioning of the drying area, changes were achieved in the company's production process; related to the time invested for each production cycle, it went from 13.05 to 9.6 hours of standard time, this time would be improved if the production process was linear. In addition, the process was improved in terms of quality and yield. In terms of quality, defective wood decreased from 34% to 23.3%, in other words, in terms of wood categories, the presence of grade A wood improved from 45.6% to 58.2%. In relation to yield, the yield increased from 229.7 to 244.2 pt/m^3 . Statistically the changes observed in yield and productivity are not significant ($p>0.433$ and $p>0.623$ respectively) (Annex 4.17 and 4.18).

The Sigma level for the lumber production process went from 2.4 to 2.7, indicating that 71.4% of the lumber was produced under adequate conditions, although it is necessary to establish improvements that will lead to a better Sigma level, which should be between 3 and 6 standard deviations (Falco, 2006).

3. COMMENT

It is necessary to highlight that the first three causes in each of the areas are related, this generates a high percentage in the low quality and yields in the wood. These aspects should be taken into account when making decisions for the improvement of production processes, so that combined improvements can be generated and implemented in the process.

The results obtained in comparison with the two scenarios described above in terms of the percentage in categories A and B, show significant differences according to the analysis of proportions ($p<0.007$). Referring to the number of defects, the differences are not significant ($p>0.695$). The moisture contents sampled showed significant differences ($p<0.000$) according to the analysis of variance.

The most difficult defects to control in this first evaluation are knots. These defects should be avoided through quality control in the forest and in the primary sawmilling process through the cutting diagram.

CONCLUSIONS

- Prior to identifying improvements, it was identified that there was no adequate quality control or wood classification in the forest. In addition, the operators do not have knowledge of the dimensions suggested by the market. This means that the raw material arriving at the company is not entirely of high quality, which leads to a decrease in yield and low quality lumber. In the yard and primary sawmilling process, it was identified that the company does not have adequate capacities in terms of cutting diagrams, ongoing preventive maintenance of the main saw, and the roundwood is kept for several days directly on the ground, which causes wood deterioration due to pests and wood fungi. The production flow does not follow a linear process, which increases production time and cost. In the drying and storage area, there is little training in classification, storage, and drying and preservation processes, and the infrastructure of the wood drying and storage area does not keep up with the volumes produced, which exposes the sawn timber to defects caused by fungi, bending, cracking, and others.

- After the process improvement was applied, it was observed that the variables of: Standard time, quality, yield, process capability index and Six Sigma level, showed positive changes, however, it is still not possible to achieve the desired changes in the process.

- The use of continuous improvement methodology such as Six Sigma with its DMAMC procedure, sets the standard for solving the problem, but success depends on the correct identification of the problem and its causes.

RECOMMENDATIONS

- Train the cooperative's personnel in quality control in both the forest and the yard. Bearing in mind that the quality of a product depends on the quality of the raw material, this also has an impact on yields, reduced time and costs.

- The relocation of the company's existing machinery should be analyzed and taken into account in order to achieve a linear flow and also consider the expansion of production capacity with the installation of an edger, resaw and trimmer, which would increase production volumes and would not overuse the main saw.

- Implement preventive and corrective maintenance plans for existing machinery, since the useful life of the company's assets depends on this.

- In order to improve the production process and achieve the desired Sigma level, continuous improvement and process control must be applied to lumber production at the cooperative.

BIBLIOGRAPHIES .

or, E., Aparicio, E., Calatayud, M & Rojas, D. (2014). *Stock reduction in spare parts warehouses in a company that manufactures corrugated cardboard boxes, applying six sigma methodology.* Lima, Peru: Universidad Peruana De Ciencias Aplicadas.

Abraham, J.C. (*n.d.*) *Manual de tiempos y movimientos Ingeniería de métodos* (1ra ed.) Tegucigalpa, Honduras. Editorial Universitaria.

Municipality of Yamaranguila (2010). *Municipal report MDG monitoring of the millennium goals.* Yamaraguila, Intibuca.

Municipality of Yamaranguila (2010,2014). *Summary of use by POA.* Yamaranguila, Intibuca: OFM.

Barbosa, J & Chancusig, W. (2012). *Improvement of the chip drying process for the manufacture of chipboard in the company Aglomerados Cotopaxi S.A, with the application of the six sigma methodology.* Quito, Ecuador: Escuela Politécnica Nacional.

Brown, N & Bethel, J. (1990). *The wood industry.* Mexico City: LIMUSA, S.A. de C. V.

Honduran Forestry Development Corporation (1991). *Manual de clasificación de la madera hondureña de dimensión.* (1ª ed.). Tegucigalpa, Honduras.

Deming, E. (1986). *"OUT OF THE CRISIS" Quality, productivity and competititve position. Cambridge:* Cambridge University Press. retrieved April 16, 2015from:

http://books.google.es/books?id=LA15eDlOPgoC&printsec=frontcover&hl=es&source=gbs_ge_summary_r&cad=0#v=onepage&q&f=false

De La Roca, C. (1994). *Manual de prácticas de ingeniería de métodos*. 1ra ed. Universidad Rafael Landívar. Guatemala.

Domínguez, D. J. (2005). *Quality Control Presentation for the University of the Americas*, Puebla, Mexico.

Dubón, P. (1996). *Comparative evaluation between the traditional manual sawmilling system and chainsaw sawmilling with frame on the north coast of Honduras*. M.Sc. Thesis. Turrialba, Costa Rica. Tropical Agricultural Research and Higher Education Center (CATIE). 123p.

EL PALIZAL. (n.d). *Bylaws of the Cooperative.* Yamaranguila, Honduras.

Falco, A. (2006). *Statistical Process Control*. Pontifical University. Madrid.

Flores, J. (*n.d.*). *Community forestry: a development option for Honduras*. USAID/ProParque. Tegucigalpa, Honduras.

Fraile, F. G., Barrio, J. F. V., & Monzón, M. T. (2003). *Six sigma*. FC Editorial. Madrid

García, R. F. (2010). *La mejora de la productividad en la pequeña y mediana empresa*. Editorial Club Universitario. España.

Gutiérrez, H. (2010). *Total quality and productivity*. 3rd ed. Editorial McGraw-Hill. México.

Grundy, I & Breton, G. L. (1998). *The SAFIRE MITI Program - A new approach to natural resource management in communal areas in Zimbabwe*. Zimbabwe.

Harry, M. J., & Stewart, R. (1988). *Six sigma mechanical design tolerancing*. Schaumburg, IL 60196-1097: Motorola University Press.

Institute of Forest Conservation and Development Protected Areas and Wildlife (2013). *Forestry statistical yearbook* (Vol. 28). Tegucigalpa; Honduras.

National Institute of Statistics (2001). *XVI population and housing census.* Tegucigalpa.

Forestry Law (2007). *Forestry, Protected Areas and Wildlife Law.* Tegucigalpa, M.D.C, Honduras.

Martinez, Y. (2011). *Six Sigma Application In The Production Process Of Poles In The Yodeco Company In Honduras.* Thesis work in forestry engineering with the academic degree of licenciatura, Escuela Nacional de Ciencias Forestales ESNACIFOR, Siguatepeque, Honduras.

Martínez, R. (2013). *Relationship between quality and productivity in SMEs in the service sector.* Publications in Science and Technology, 7(1), 85-102 p.

Manivannan, S. (2007). *Introduction to six sigma.* METALFORMING, 51-52 p. Retrieved March 29, 2015 from: http://mexico.pma.org

Marquez, A & Pérez, L. (2007). *Study of the intervening factors in the innovative process of metal-mechanical SMEs.* Tachira State, Venezuela.

Moori, G. (n.d.). *Measurement of work: Normal time, standard time.* Peru.

Standard, I.S.O. (2008). 9001:2008. *quality management system.* (4ª ed.). Geneva; Switzerland.

I.S.O. 9000 Standard. Accessed March 26, 2008. Available at http://www.adrformacion.com/cursos/calidad/leccion1/tutorial2.html

Obregón, M., De La Torre, A., Díaz, V., Rodríguez, L., Fernández, D., Domínguez, V., Piñero, A., *et al* (2008). *Training and development of human resources.* Ministry of Public Health. Havana, Cuba.

International Tropical Timber Organization (2000). *Criteria and indicators for the sustainable management of natural tropical forests.* ITTO Forest Policy

Series No. 7. Yokohama. Japan.

Parameda, C., Brenes, E., & Figueroa, L. (1998). *The Wood Industry in Honduras:* Competitiveness Conditions. *CEN Working Paper, 534.*

Management Plan (2008). *Yamaranguila ejido site.* Reg. BE-L2-002-98-III From ICF.

Pries, K. (2006). *Six Sigma for the Next Millennium, ASQ Quality Press,* Milwaukee, Wisconsin, USA.

Salazar, M. (2011). *Systematization of the experience of the PALISAL Cooperative. Yamaranguila,* Honduras.

Ulloa, U. (2001). *Plan de manejo forestal 2002-2006. Yamaranguila, Intibuca:* MAFOR/LENCAFOR.

Vasquez, Y. (2012). *Study of profitability and yield using three types of saws in the production of sawn timber of Pinus oocarpa SCHIEDE at the National School of Forestry Sciences (ESNACIFOR), Siguatepeque, Honduras.* Thesis work in forestry engineering with the academic degree of licenciatura, ESNACIFOR, Siguatepeque, Honduras.

Vootukuru, A. S. (2008). *DMARC: A Framework for the Integration of DMAIC and DMADV.* University of Central Florida. Orlando, Florida.

ANNEXES

1.1   *Cause* and *effect diagram of the main problems found in the forest.*

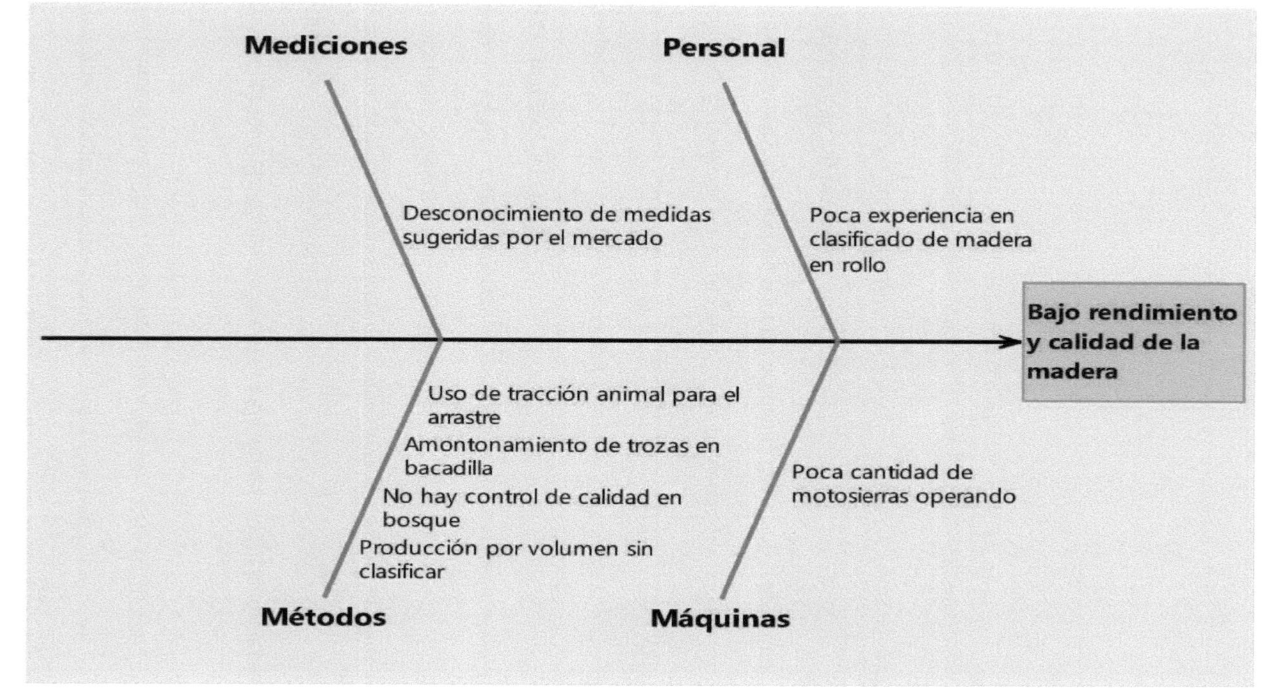

4.2   *Cause* and *effect diagram of the main problems encountered in the yard and primary sawmill.*

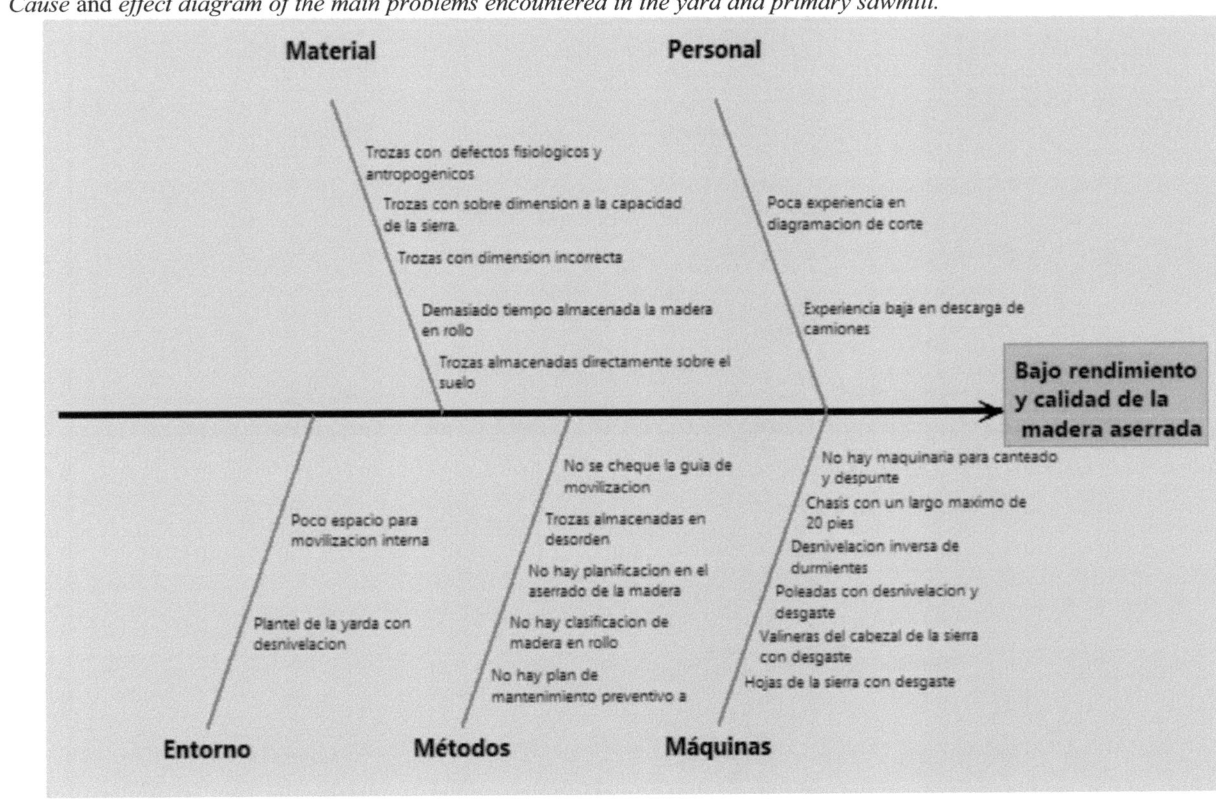

4.3   *Cause* and *effect diagram of the main problems encountered in the drying* and *storage area.*

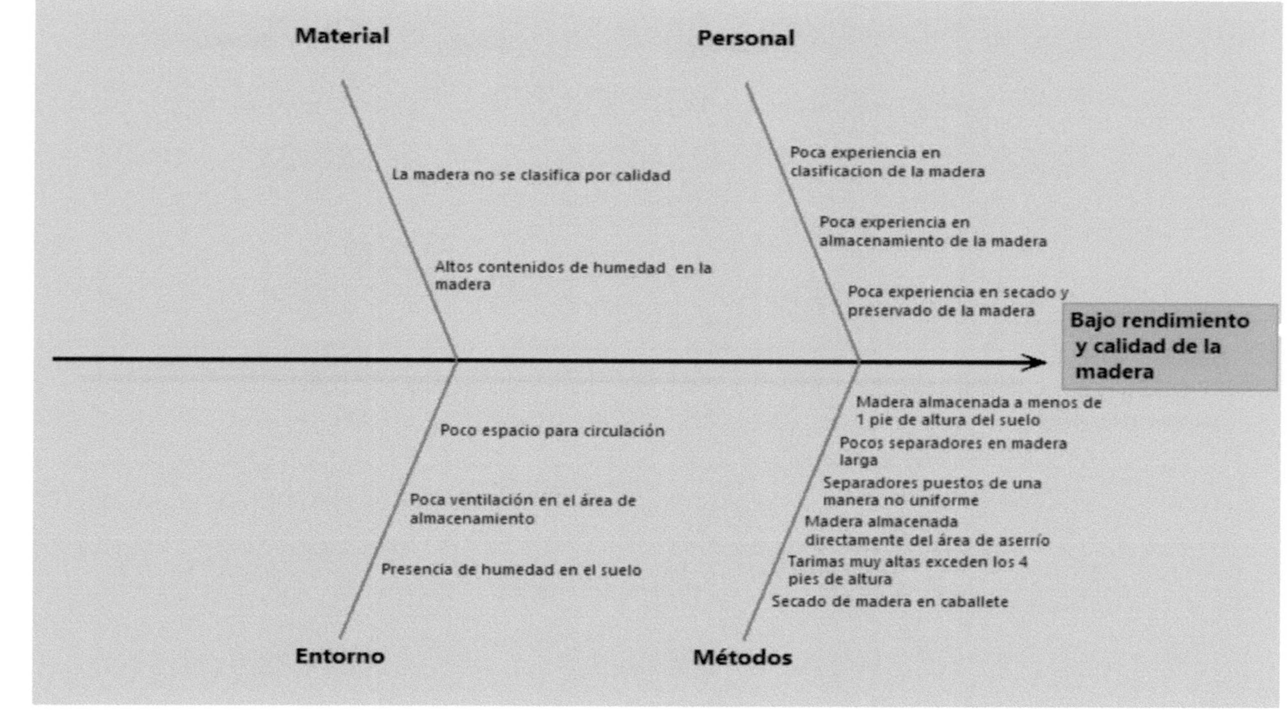

4.4   *Form for time* and *motion data recording*

FORMAT FOR TIME AND MOTION STUDY

Observer:Operator:

Date:Start time:End time:

Description of the activity	Cycles										Summary				
	1	2	3	4	5	6	7	8	9	10	XT	T.P	T.O	Faith	TN
											Minutes per cycle				
	Standard time: _______________														

Source: Martinez, 2011

Time summation (XT), average time (T.P), qualification factor (Fe), normal time (TN), average operation time (T.O).

4.5 Performance Control Form

CICL or DATE: **SEWER:** **CUBICATOR:**

HOURS WORKED: **REMARKS:**

SPECIES:
Pine

Piece	ROLL WOOD (Raw Material)				SAWN LUMBER											
	Diameter (in)	Length (Feet)	Vol. (P3)	Vol. (M3)	No. Parts	Thickness (in)	Width (in)	Length (Foot)	Vol. (Pt)	Detail	No. Parts	Thickness (in)	Width (in)	Length (Foot)	Vol. (Pt)	Detail
1																
2																
3																
4																
5																
6																
7																
8																
9																
10																
11																
12																
13																
TOTAL				0.0	TOTAL				0.0		TOTAL					
TOTAL IN M3:	PRODUCTION IN PT:				YIELD PT/M3:			YIELD IN %:			PRODUCTIVITY PT/HOUR:					

4.6 Quality control form

PRODUCTION AND QUALITY CONTROL OF SAWN TIMBER

Date: **Sawyer:**

Species: **Shift:** **Remarks:**

ROUND WOOD (Raw Materials)		SAWN LUMBER																				Quality		
Log			Measures				Defects in (%)																	
N o / Measures	Nº Pieza	Esp (in.	Width (in)	Length (foot)	Vol (Pt)	Knots	Bags of	Between hulls	Rot	Fendas	Measures not respected	Humidity	Stains due to	Wood of	Deviation from llflbra	Striped wood	Gems	Cracks	Medulla	Crack	A	B	No. Defects
Diam < (in)	1																						
	2																						
Length (foot)	3																						
	4																						
Vol (m3)	5																						
	6																						
Defects (%)	7																						
Knots	8																						
Burning	9																						
Sprain	10																						
Crack	11																						
Rot	12																						
Others	13																						
	14																						
	15																						

71

4.7   *Westinghouse System Rating Table*

	Ability			Effort	
+0.15	To	Super Handy	+0.13	To	Excessive
+0.13	A2	Super Handy	+0.12	To	Excessive
+0.11	Bl	Excellent	+0.10	Bl	Excellent
+0.08	B2	Excellent	+0.08	B2	Excellent
+0.06	C1	Good	+0.05	C1	Good
+0.03	C2	Good	+0.02	C2	Good
0.00	D	Average	0.00	D	Average
-0.05	The	Regular	-0.04	The	Regular
-0.10	I E2	Regular	-0.08	E2	Regular
-0.16	> F1	Poor	-0.12	F1	Poor
-0.22	F2	Poor	-0.17	F2	Poor
	Conditions			Consistency	
+0.04	A	Ideal	+0.04	A	Perfect
+0.04	B	Excellent	+0.03	B	Excellent
+0.02	C	Good	+0.00	C	Good
0.00	D	Average	0.00	D	Average
-0.03	E	Regular	-0.02	E	Regular
-0.07	F	Poor	-0.04	F	Poor

4.8 *Summary of quality control field data in percentage of defects in pre-improved sawn lumber.*

Sample	No. Pieces	Knots	Bags of resin	Between hulls	Rot	Fendas	Fungal stains	Wood striped	Gems	Cracks	Marrow
1	6	30	0	0	0	0	0	0	0	0	0
2	6	16	28	0	0	0	0	0	0	0	0
3	9	99	78	7	0	0	0	0	107	0	0
4	5	24	0	0	0	0	0	0	0	0	0
5	7	55	55	130	0	0	0	0	0	0	0
6	6	18	0	0	0	0	0	0	81	0	0
7	7	119	33	20	0	0	0	0	33	0	0
8	9	30	110	0	0	0	0	0	140	0	0
9	9	148	0	0	0	0	0	0	57	0	0
10	10	35	50	10	0	0	0	0	7	63	0
11	9	0	46	7	0	0	0	0	83	0	0
12	13	0	154	0	0	0	0	0	178	299	0
13	7	7	108	0	0	0	0	0	57	0	0
14	19	168	199.5	0	0	0	0	0	164	29	0
15	11	45	36	0	0	0	0	0	36	0	0
16	7	62	0	0	0	7	0	0	19.5	0	14
17	4	20	15	0	0	0	0	0	30	0	45
18	6	73	0	0	0	0	0	0	18	0	100
19	5	39	0	0	0	0	0	0	11	0	0
20	8	27	155	0	0	0	0	0	20	0	0
21	6	10	15	0	0	0	0	0	40	0	0
22	11	122	261	0	0	0	0	0	41	0	0
23	4	55	0	0	0	0	0	0	0	17	10
24	7	45	7	0	0	0	0	0	48	0	7
25	11	80	0	0	0	0	0	0	25	0	274
26	7	48	421	0	0	0	0	0	0	0	0
27	8	14	175	0	0	0	0	0	56	0	0
28	7	49	137	0	0	0	0	0	17	17	0
29	6	10	30	0	0	0	0	0	45	0	0
30	16	163	69	8	0	0	0	0	191	0	200
31	10	109	0	0	0	0	0	0	21	0	46

32	6	43	0	0	0	0	0	0	0	0	100
33	8	170	10	0	0	0	0	0	40	0	0
34	8	98	0	0	0	0	0	0	42	28	0
35	13	337	8	0	0	0	0	0	0	0	16
36	6	154	0	0	0	0	0	0	68	0	86
37	7	66	0	0	0	0	0	0	10	10	277
38	9	36	25	0	0	0	0	0	16	0	12
39	9	96	0	0	0	0	0	0	41	0	8
40	6	44	178	0	0	0	0	0	70	0	0
41	8	8	249	7	0	0	0	0	51	0	0
42	5	43	33	0	0	0	0	60	143	0	0
43	12	435	113	0	0	15	0	48	38	0	437
44	6	212	0	0	0	0	0	29	0	0	71
45	5	192	0	12	0	0	0	88	6	0	232
46	5	241	0	0	0	0	0	0	14	0	200
47	5	98	119	0	0	0	0	0	7	0	140
48	7	170	140	0	0	0	0	44	80	0	30
49	11	279	24	0	0	0	0	96	0	0	199
50	25	109	154	0	0	5	0	0	0	46	0
51	25	100	345	0	0	70	300	0	50	28	0
52	25	258	510	0	0	0	615	0	147	0	0
		10.5	8.7	0.4	0.0	0.2	2.0	0.8	5.0	1.1	5.3
							Parts	469.0			34.0

4.9   *Summary of quality control field data on number of defects in pre-improved sawn lumber*

QUALITY OF SAWN TIMBER FROM THE EL PALISAL COOPERATIVE (M					pre-improvement edition June 18 to 21, 2015)				
Number	Classif. A	Classif. B	No. Defects	Humidity % Humidity	TP Number	Classif. A	Classif. B	No. Defects	Humidity % Humidity
6	6	0	7	29.2	7	7	0	23	27.6
6	6	0	8	28.5	11	6	5	37	27.3
9	3	6	37	28.2	7	1	6	40	27.2
5	5	0	9	26.1	8	1	7	16	27.9
7	3	4	TI	27.8	7	2	5	38	27.5
6	5	1	10	27.3	6	5	1	11	27.5
7	2	5	29	28.9	18	7	11	34	25.1
9	3	6	25	28.4	10	9	1	17	32.7
9	4	5	31	28.1	6	4	2	9	23.5
10	6	4	20	28	8	4	4	27	23.9
9	7	2	18	27.9	8	5	3	15	28.1
13	3	10	46	25.7	13	6	7	43	28.7
7	5	2	16	27.9	5	0	5	30	29.1
19	6	13	54	27.3	5	1	4	19	29.2
11	9	2	16	27.4	6	0	6	31	TI
7	5	2	16	23.6	7	2	5	34	TI
5	1	4	13	29.4	9	7	2	18	29.9
12	2	10	64	28.4	9	8	1	19	23.4
4	2	2	18	25.5	7	0	7	35	28.3
6	4	2	33	26.4	11	0	11	36	30.4
5	5	0	17	25	6	1	5	20	24.5
8	4	4	25	29	8	2	6	10	24.9
6	5	1	4	26.3	25	19	6	50	22.3
11	1	10	40	27.7	25	9	16	48	18.9
4	2	2	19	25.8	25	3	22	58	23.8
6	1	5	34	29.4	**212**	**105**	**107**	**662**	**16.86**

				0	5	26	28.5			49.5%	50.5%

4.10   Summary of quality control field data in percentage of defects in post improvement sawn lumber.

Sample	No. Parts	Knots	Resin bags	Fendas	Fiber deflection	Gems	Marrow
1	7	105	0	0	0	46	30
2	11	215	88	15	0	40	63
3	7	167	15	0	0	20	25
4	8	145	10	0	0	40	38
5	10	108	48	0	0	30	200
6	10	60	60	0	5	38	70
7	8	134	133	0	0	24	15
8	7	78	17	0	0	15	0
9	7	32	24	0	0	20	0
10	7	70	20	0	0	32	13
11	9	84	28	0	0	38	6
12	4	20	0	0	0	5	0
13	11	91	0	0	0	44	0
14	6	31	36	0	0	21	0
15	7	61	71	0	0	24	0
16	8	30	40	0	0	25	0
17	7	85	55	0	0	25	0
	11.3		4.8	**0.1**	**0.04**	**3.6**	3.4
			Parts		134	**23**.3	

4.11   Summary of quality control field data on number of defects in post-improvement sawn lumber

QUALITY OF SAWN TIMBER COOPERATIVA EL PALISAL

					Post-improvement measurement August 01 5 to 13, 201				
Number	Classif. A	Classif. B	No. Defects	Humidity % Humidity	TP Number	Classif. A	Classif. B	No. Defects	Humidity % Humidity
7	2	5	18	27.06	9	5	4	28	21.5
11	6	5	28	23.8	4	4	0	8	23.8
7	3	4	22	27.13	11	8	3	21	27.1
8	4	4	32	22	6	4	2	15	22.9
10	4	6	29	24.4	7	4	3	18	20.2
10 6		4	29	25.3	8	6	2	21	15.4
8	3	5	22	27.2	7	4	3	26	13.8
7	5	2	21	22.6	134	78	56	383	22.8
7 6		1	19	23.5				58.2%	41.8%
7	4	3	26	19.9					

4.12 Summary of field data - time and motion study - process without improvement

TIME AND MOTION STUDY COOPERATIVE EL PALISAL

WITHOUT IMPLEMENTING PROCESS IMPROVEMENTS

Processed logs		18	13	7	14	15	5	15	18	8	11					

Sawmill/cycle	1	2	3	4	5	6	7	8	9	10		T.P	Fait	T.0	T.N	TS
Log restoration / length adjustment	21.1	24.7	19.2	7.4	14.2	6.2	11.2	20.3	5.3	10.6	140.1	14.0	1.3	14.0	17.7	24.2
log mounting	32.4	23.4	12.6	25.2	21.3	7.1	10.5	35.6	43.2	19.4	230.7	23.1	1.3	23.1	29.1	39.8
cuts	209.0	150.9	81.3	162.5	175.2	58.4	82.4	120.2	11.9	76.5	1128.3	112.8	1.3	112.8	142.2	194.7
overturns	37.1	26.8	14.4	28.8	50.4	16.8	24.3	32.6	40.8	15.4	287.4	28.7	1.3	28.7	36.2	49.6
canteo	101.9	73.6	39.6	79.2	141.0	47.0	78.2	16.0	14.0	9.9	600.4	60.0	1.3	60.0	75.6	103.6
blunt	35.1	25.4	13.7	27.3	37.8	12.6	28.4	23.0	116.2	6.8	326.3	32.6	1.3	32.6	41.1	56.3
TOTAL	436.5	324.7	180.8	330.5	439.8	148.1	234.8	247.7	231.4	138.5				271.3	341	468
Truck unloading	1	2	3	4	5	6	7	8	9	10						
Location of sleepers	2.3	1.2	1.5	8.7	5.3	4.4	2.0	6.6	2.5	3.2	37.7	3.8	1.2	3.8	4.6	6.4
Chain loosening	2.7	1.6	1.1	1.3	1.9	2.2	1.2	1.4	1.3	2.0	16.6	1.7	1.2	1.7	2.0	2.8
Cutting of lateral supports	7.2	5.4	3.0	4.0	3.5	4.3	3.5	4.1	5.3	4.1	44.4	4.4	1.2	4.4	5.5	7.5
Log unloading	28.0	28.2	20.0	22.0	23.4	25.1	21.6	22.1	26.6	22.1	239.1	23.9	1.2	23.9	29.4	40.3
TOTAL	40.16	36.38	25.58	35.91	34.06	35.99	28.31	34.27	35.67	31.46				33.78	41.55	56.92
Machine stops	1	2	3	4	5	6	7	8	9	10						
Cleaning and blade change	28.4	40.2	44.2	52.1	34.5	39.1	51.0	38.1	48.3	53.2	429.1	42.9	1.2	42.9	53.2	72.9

	1	2	3	4	5	6	7	8	9	10		T.P	Fait	T.O	T.N	TS
Flyer check	5.4	6.3	0.0	2.2	0.0	4.3	3.2	2.1	0.0	2.0	25.5	2.6	1.2	2.6	3.2	4.3
Band adjustment	2.3	3.6	0.0	0.0	2.8	3.1	0.0	1.9	0.0	0.0	13.7	1.4	1.2	1.4	1.7	2.3
TOTAL	36.12	50.11	44.21	54.31	37.3	46.5	54.18	42.07	48.31	55.18				46.83	58.0	79.
Wood storage	1	2	3	4	5	6	7	8	9	10						116.5
Selection and transport of wood	71.2	67.5	76.6	74.1	63.9	69.2	64.6	77.1	70.2	68.4	702.8	70.3	1.2	70.3	85.0	
manual cleaning of bark on boards	16.1	7.8	5.2	14.6	7.6	17.1	8.8	4.9	18.6	6.6	107.3	10.7	1.2	10.7	13.0	17.8
cutting of separators	14.5	12.6	10.0	14.0	14.6	15.4	11.5	11.2	13.0	10.7	127.3	12.7	1.2	12.7	15.4	21.1
wood stacking	12.8	11.3	10.9	16.5	19.1	13.6	12.3	9.1	14.6	20.5	140.8	14.1	1.2	14.1	17.0	23.3
TOTAL	114.6	99.2	102.6	119.2	105.1	115.3	97.2	102.3	116.3	106.2				107.8	130	178
GRAND TOTAL/minutes	627.4	510.4	353.3	539.9	616.4	345.9	414.5	426.5	431.7	331.4				459.7	571.9	783.5
GRAND TOTAL/hours	10.5	8.5	5.9	9.0	10.3	5.8	6.9	7.1	7.2	5.5				7.66	9.53	13.06

4.13 *Summary of field data time* and *motion study process with improvement*

TIME AND MOTION STUDY COOPERATIVE EL PALISAL																
APPLIED TO THE PROCESS IMPROVEMENTS																
Processed logs	6	8	8	8	6	8	9	7	6	10						
Sawmill/cycle	1	2	3	4	5	6	7	8	9	10		T.P	Fait	T.O	T.N	TS
Log restoration / length adjustment	0.0	7.2	10.2	0.0	8.5	5.2	11.2	12.2	6.3	0.0	60.5	6.1	1.3	6.1	7.6	10.4

log mounting	6.8	9.3	10.2	11.2	9.3	13.2	13.5	13.3	11.2	15.3	113.0	11.3	1.3	11.3	14.2	19.5
cuts	81.8	93.4	92.9	105.1	48.7	60.1	92.9	81.3	69.7	116.1	841.9	84.2	1.3	84.2	106.1	145.3
overturns	23.5	26.9	16.5	30.2	9.8	16.3	16.5	14.4	12.4	20.6	187.1	18.7	1.3	18.7	23.6	32.3
canteo	65.8	75.2	45.3	84.6	6.3	8.0	45.3	39.6	34.0	56.6	460.7	46.1	1.3	46.1	58.0	79.5
blunt	17.6	20.2	15.6	22.7	4.3	11.5	15.6	13.7	11.7	19.5	152.4	15.2	1.3	15.2	19.2	26.3
TOTAL	195.5	232.1	190.5	253.9	86.8	114.2	194.9	174.4	145.1	228.1				181.6	228.8	313.4
Truck unloading	1	2	3	4	5	6	7	8	9	10						
Location of sleepers	1.9	1.5	1.5	4.3	3.2	5.4	2.3	4.9	3.3	3.2	31.2	3.1	1.2	3.1	3.8	5.3
Chain loosening	1.9	1.6	1.3	1.2	2.1	2.4	1.2	1.3	1.3	2.0	16.1	1.6	1.2	1.6	2.0	2.7
Cutting of lateral supports	8.1	5.2	4.3	3.6	3.3	4.0	3.6	4.2	4.4	4.3	44.7	4.5	1.2	4.5	5.5	7.5
Log unloading	26.6	30.2	18.2	22.1	25.1	25.2	20.3	23.5	20.2	23.3	234.4	23.4	1.2	21.6	26.6	36.4
TOTAL	38.4	38.5	7.0	31.1	33.6	36.9	27.3	33.8	29.0	32.6				30.8	37.9	51.9
Machine stops	1	2	3	4	5	6	7	8	9	10						
Cleaning and blade change	15.3	21.1	18.3	25.1	26.2	18.5	24.2	30.0	19.2	23.2	220.8	22.1	1.2	22.1	27.4	37.5
Flyer check	0.0	4.3	2.2	2.4	1.5	5.3	4.1	0.0	0.0	2.3	21.9	2.2	1.2	2.2	2.7	3.7
Band adjustment	0.0	4.3	0.0	0.0	2.2	0.0	0.0	3.2	0.0	0.0	9.6	1.0	1.2	1.0	1.2	1.6
TOTAL	15.3	29.6	20.4	27.5	29.8	23.8	28.3	33.2	19.2	25.4				25.2	31.3	42.8
Wood storage	1	2	3	4	5	6	7	8	9	10						
Selection and transport of wood	65.2	68.3	82.2	70.2	50.5	75.3	60.3	50.4	62.3	78.2	662.6	66.3	1.2	66.3	80.2	109.8

manual cleaning of bark on boards	14.9	7.2	4.8	13.5	7.0
cutting of separators	13.4	11.7	9.2	12.9	13.5
wood stacking	11.8	10.5	10.1	15.3	17.7
TOTAL	114.6	99.2	102.7	119.2	105.1
GRAND TOTAL/minutes	363.81	399.42	320.63	431.58	255.30
GRAND TOTAL/hours	6.1	6.7	5.3	7.2	4.3

15.8	8.1	4.6	17.2	6.1	99.3	9.9	1.2	9.9	12.0	16.5
14.3	10.6	10.4	12.0	9.9	117.8	11.8	1.2	11.8	14.3	19.5
12.6	11.4	8.4	13.5	19.0	130.3	13.0	1.2	13.0	15.8	21.6
115.3	97.2	102.4	116.3	106.2				101.0	122.2	167.4
290.21	347.64	343.73	309.54	392.29				338.6	420.16	575.6
4.8	5.8	5.7	5.2	6.5					7.0	9.6

4.14 Summary of field data yield study

SUMMARY OF PRODUCTION AND YIELDS EL PALISAL COOPERATIVE (PRE-IMPROVEMENT)

Sampling from June 15 to 24, 2015.

CYCLE	M3 PROCESSED	PT PRODUCED	YIELD PT/M3	YIELD IN %	PRODUCTIVITY PT/HOUR	HOURS
1	6.004	1569.9	261.47	61.7	149.51	10.5
2	4.200	851.0	202.62	47.8	100.12	8.5
3	2.450	725.7	296.22	69.9	122.99	5.9
4	4.860	1045.6	215.16	50.7	116.18	9
5	4.602	848.1	184.29	43.5	82.34	10.3
6	2.065	629.0	304.66	71.9	108.45	5.8
7	5.140	1142.3	222.24	52.4	165.56	6.9
8	5.245	1075.3	205.02	48.4	151.46	7.1
9	2.167	447.3	206.48	48.7	62.13	7.2
10	2.809	558.8	198.92	46.9	101.61	5.5
PROM.		**8893.1**	**229.7**	**54.2**	**116.03**	**76.7**

SUMMARY OF PRODUCTION AND YIELDS AT EL PALISAL COOPERATIVE (POST IMPROVEMENT)

Sampling from August 01 to 13, 2015.

CYCLE	M3 PROCESSED	PT PRODUCED	YIELD PT/M3	YIELD IN %	PRODUCTIVITY PT/HOUR	HOURS
1	3.741	972.3	259.90	61.3	138.9	7

2	4.511	1033.33	229.07	54.0	103.3	10
3	2.911	713.67	245.13	57.8	101.9	7
4	3.812	824.33	216.25	51.0	137.4	6
5	2.203	716.2	325.10	76.7	119.4	6
6	3.039	899	295.82	69.8	128.4	7
7	3.908	859.2	219.86	51.9	107.4	8
8	3.223	698.5	216.72	51.1	116.4	6
9	2.727	606.08	222.25	52.4	151.5	4
10	4.274	907.3	212.30	50.1	113.4	8
PROM.		**8229.9**	**244.2**	**57.6**	**121.8**	**69.0**

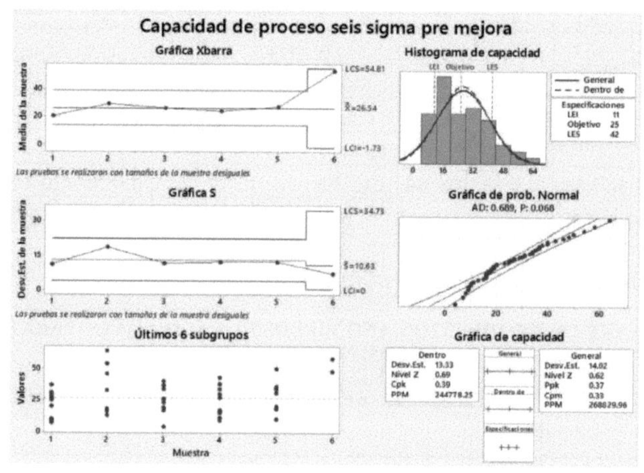

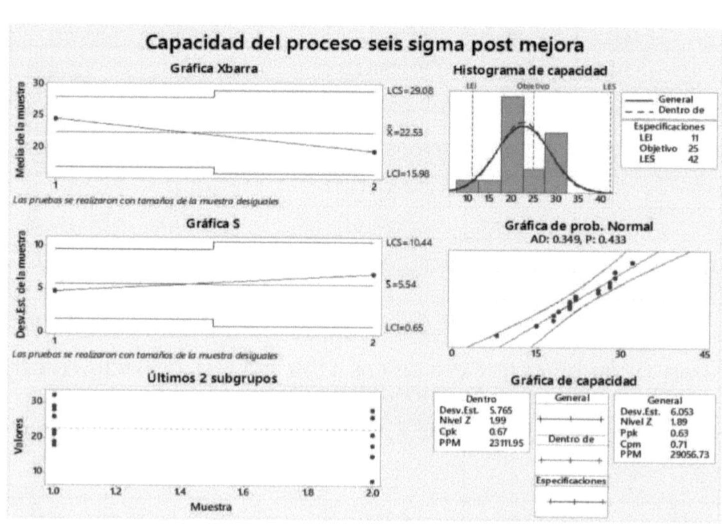

4.15 Process capability graph without improvement

4.15 Process capability graph with improvement

4.16 Analysis of Variance

4.17 Analysis of Variance for Yields

	MC	
Source	GL SC Adj. Adj. F-value p-value	
SCENARIO	1105610560.	640.433
Error	18295271640	
Total	1930583	

4.18 Analysis of variance for productivity

Analysis of Variance

Source	GL	SC Adjust.	MC Adjust.	F-value	p-value
SCENARIO	1	166.2	166.2	0.25	0.623
Error	18	11948.7	663.8		
Total	19	12114.9			

4.19 Analysis of variance for operation time

Analysis of Variance

Source	GL	SC Adjust.	MC Adjust.	F-value	p-value
SCENARIO	1	18.22	18.220	8.96	0.008
Error	18	36.62	2.034		
Total	19	54.84			

4.20 Ratio analysis for wood quality by category

Test and CI for two ratios (wood grading category A/0 at 20% defects)

Sample	X	N	Sample p
1	78	134	0.582090
2	214	469	0.456290

Difference = p (1) - p (2)
Estimated difference: 0.125800
Lower limit 95% of difference: 0.0461581
Test for difference = 0 vs. > 0: Z = 2.60 p-value = 0.005

Fisher's exact test: p-value = 0.007

Test and CI for two ratios (classification of wood in category B/20 at 100% defects)

Sample	X	N	Sample p
1	255	469	0.543710
2	56	134	0.417910

Difference = p (1) - p (2)
Estimated difference: 0.125800
Lower limit 95% of difference: 0.0461581
Test for difference = 0 vs. > 0: Z = 2.60 p-value = 0.005
Fisher's exact test: p-value = 0.007

4.21 Analysis of variance for defects

Analysis of Variance

Source	GL	SC Adjust.	MC Adjust.	F-value	p-value
Scenario	1	27.4	27.37	0.15	0.695
Error	67	11844.9	176.79		
Total	68	11872.3			

4.22 Analysis of Variance for Wood Moisture

Analysis of Variance

Source	GL	SC Adjust.	MC Adjust.	F-value	p-value
Scenario	1	233.2	233.152	30.20	0.000
Error	67	517.2	7.719		
Total	68	750.3			

EL PALISAL

TRANSCRIPCION DE PUNTO DE ACTA

El suscrito presidente de la junta central de la directiva de la Cooperativa Agroforestal El Palisal de Yamaranguila Intibucá, en uso de sus facultades que la ley le confiere transcribe el punto de acta No. 132.5 ubicado en el libro de actas de la junta directiva de la cooperativa agroforestal, del día 04 de julio del año 2015.

La reunión se llevó a cabo en las instalaciones del plantel de la cooperativa agroforestal El Palisal, en la que se discutieron diferentes puntos enmarcados en la agenda de reunión, dentro de ellos se dio apertura a la participación como punto número cinco (5) de la agenda al Dasónomo José Aquimix Pérez Rodríguez tesista de la Escuela Nacional en Ciencias Forestales (ESNACIFOR). El tesista informo de forma detallada a la junta directiva en base a análisis anteriores llevados a cabo en el plantel de la sierra las diferentes causas que generan el bajo rendimiento y calidad de la madera y en conjunto se determinó aplicar una mejora combina tomando en cuenta los siguientes aspectos:

1. Clasificar la madera en rollo puesta en bacadía, para evitar la llegada de madera con defectos al plantel de la sierra.

2. Capacitar a los operarios en el tema de diagramación de corte de madera en rollo.

3. Mejorar las condiciones del área de secado para evitar defectos como; mancha azul, agrietamiento de la madera y otros.

4. Dar manteniendo preventivo a la sierra Wood Mizer, además del cambio de algunas piezas dañadas de la sierra.

Dado en el municipio de Yamaranguila Intibucá, a los 28 días del mes de Septiembre del año 2015.

José German Pérez Gómez
Presidente de la junta directiva central

4.23 Implementation of the improvement

CATIE | Proyecto Finnfor
Bosques y Manejo Forestal
en América Central

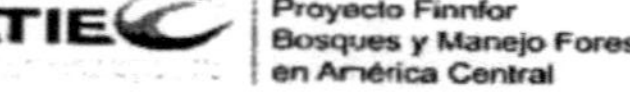

MINISTRY FOR FOREIGN
AFFAIRS OF FINLAND

Proyecto Finnfor II
"Fortalecimiento de capacidades en el mantenimiento preventivo de máquinas"

Fechas: 28 junio del año 2015

Lugar: Planta de procesamiento de El Palisal, Intibucá, Honduras.

No	Nombre y apellido	Sexo		Institución	Teléfono	Correo electrónico	Firma
		H	M				
01	CARLOS Lorenzo Fores	H	M	El PALISAL	—	—	
02	Giacinto Lopez Reyes	H		El Palisal	94979445	Socio	
03	Evelio Arriaga R.	H		El Palisal	96945936	Secretario Palisal	
04	José Bernardino Sánchez	H		El Palisal	94572271	Ayudante Aserrio	
05	Bernabé Vásquez	H		El Palisal		Bioyero en cerro	
06	Rafael Sánchez Perez	H		Palisal	97715473	Tesorero Palisal	
07	José Aguirre Rodríguez	H		Tesista/Palisal	32402144	Tesista en Palisal	
08	Wilson Guerra Arevalo	H		CATIE/Finnfor	—	wilson.guerra@catie.ac.cr	
09	Wilson M. Vásquez Meza	H		Instructor ind	9905-7475	w.krakeo@hotmail.com	
10							
11							
12							
13							
14							
15							
16							
17							
18							
19							

Wilson Guerra Arevalo
Tesista CATIE/Finnfor

Das. Wilson Miguel Vásquez Meza
CAPACITADOR.

Proyecto Finnfor II

"Fortalecimiento de capacidades en la diagramación del corte de madera de pino en rollo"

Fechas: Sábado 25 julio del año 2015

Lugar: Planta de procesamiento de El Palisal, Intibucá, Honduras.

No	Nombre y apellido	Sexo		Institución	Teléfono	Correo electrónico	Firma
01	Jose Esteban Perez Gomez	(H)	M	Palisal	94692276	vocal 1 Junta Directiva	
02	Julian Reyes	H		PALISAL	58806602	Socio	
03	Pedro Sanchez 12.	H		PALISAL	98790072	Socio	
04	Lanito Sanchez		M	INVITADA	98533890	INVITADA	
05	Fausto Lopez	H		PALISAL	94979445	Socio	
06	Rafael Sanchez	H		PALISAL	97215443	Tesorero Junta Directiva	
07	Evelio Arriaga Morales	H		PALISAL	96945936	Secretario Junta Directiva	
08	Miguel Sanchez	H		PALIZAL	—	Socio	
09	FAUSTINO RODRIGUEZ	H		Palizal	33854897	Socio	
10	WILSON F. GUERRA AREVALO	H		CATIE/finnfor	96455933	Tesista/Organizador	
11	Jose Aquimis Rodriguez	✓		Escuela/Tesista	32402149	Tesista/Ejecutor	
12							
13							
14							
15							
16							
17							
18							
19							

Printed by Books on Demand GmbH, Norderstedt / Germany